how to
get expelled
from school

A guide to climate change for pupils, parents & punters

Ian Plimer

Foreword by Dr Václav Klaus, President of the Czech Republic

connorcourt
PUBLISHING

Published in 2011 by Connor Court Publishing Pty Ltd.

Connor Court Publishing Pty Ltd.
PO Box 1
Ballan VIC 3342
sales@connorcourt.com
www.connorcourt.com

ISBN: 9781921421808 (pbk.)

Front cover design: Ian James

Printed and designed in Australia.

This book is dedicated to Jára Cimrman, an international polymath recognised some 50 years after his death and then only by highly creative people.

Best wishes,

Ian Peine

CONTENTS

FOREWORD

In the last decade, we have been authoritatively told and forced to accept that the Earth has been warming up. Some of us are not ready to accept that. Before fully acceding to the global warming alarmism, we want to know why, how much and how relevant (if not dangerous) the warming is or could be. Many politicians, journalists and activists among scientists pretend to know all the answers. According to them, it is caused by man and his irresponsible activities (by emitting CO_2), as a result of which the warming will be unusually high and will have disastrous consequences. Moreover, the "science is settled", they claim, which is not at all true.

There are many serious opponents, "skeptics", "deniers", people whose views are different. One of them is Professor Ian Plimer.

Professor Plimer is a well-respected scientist, both in Australia and worldwide. He is a Professor at The University of Adelaide and at The University of Melbourne. He is Australia's best-known geologist. In his almost half a century lasting academic career, he has published more than one hundred scientific books and papers.

In the last couple of years, he started addressing a wider audience. He reacted to the fact that one of his scientific disciplines – earth science- had been stolen by non-scientists who had been assisted by scientists who prefer TV performances, celebrity status, fame and luxurious life to the simple, solemn, financially less rewarding life of a scholar. He also understood the consequences of all that. He understood that climatology had converted into a servant of global warming doctrine that is not a science, but an ideology, if not religion, based on the idea that man, by emitting CO_2, controls the climate. This simple doctrine says that rising, man-made CO_2 emissions drive up the temperature on Earth.

As a serious scientist who studied global climate fluctuations in not

1

only hundreds or thousands, but millions of years, Professor Plimer knows that there are other mechanisms behind the global climate and that the role of man-made CO_2 emissions is – if any – very small, almost negligible.

There is no doubt that the scientific debate about this issue is still open but the public but the public and political debate was prematurely closed and declared "decided". This is the reason why Professor Plimer moved from science to the public debate, to the dispute with the global warming doctrine and its adherents. His book *Heaven and Earth,* published in 2009, became a worldwide success and influenced public debate in many countries, most notably in "his" own Australia. I learnt a lot from it and was honoured to be asked to endorse it. I did it by saying – among other things – that "If I'd had the chance to read Professor Plimer's book in advance, my own book on the subject of global warming would have been better."[1]

As indicated by its subtitle Global warming: The missing science, *Heaven and Earth* is a very professional and sophisticated text which asks for studying, not just reading, and is, therefore, not easily accessible to everyone. That's why Professor Plimer wrote a sequel to it with a rather unexpected – and for some people perhaps confusing – title *How to get expelled from school.* The title suggests that the book is not a scientific text, that is – as its subtitle indicates – A guide to climate change for pupils, parents and punter. This form of argumentation is a much needed contribution to the worldwide debate about this issue. I am only afraid that – with all its simplicity – it's not a text for children. Even this book must not only be read but studied.

Investing in reading and studying this book will bring very high returns to everyone. We have to thank Professor Plimer for his non-traditional approach and for his scholarly achievement.

Dr Václav Klaus, President of the Czech Republic
20th September 2011

---

1 *Blue Planet in Green Shackles. What is endangered: Climate or Freedom?* Competitive Enterprise Institute, Washington D.C., 2007.

PREAMBLE FOR PUPILS, PARENTS AND PUNTERS

This book is deliberately seditious. There is no climate emergency. We are all environmentalists and we all want a better world for ourselves and the next generation. Pollution kills and no one wants pollution of the atmosphere, waterways and soil. By contrast, carbon dioxide is plant food. It is not a pollutant. How can it be dangerous if we breathe it in and breathe it out? Without sunlight, water and carbon dioxide there would be no life on Earth. To call carbon dioxide a pollutant is fraud.

Previous generations feared the cold. They knew that far more people die in cold weather than warm weather. The current generation of school pupils may be the first generation to fear warmth, something that humans have been seeking for tens of thousands of years.

We cannot afford to wastefully allocate our sparse resources to ventures that give us false hope and address false problems. We cannot be guided by activists who cherry-pick information to suit their ideology. We must be very wary of banks, corporations, traders and lobbyists seeking huge subsidies for schemes to reduce our emissions using our money. We must be very wary of those self-appointed saviours who claim that they know what is good for us. They don't. Carbon dioxide is a minor greenhouse gas. At least 95% of the greenhouse effect in the atmosphere is due to water vapour.

What we do know is that climate always changes, carbon dioxide has a decreasing warming effect with increasing concentration and that humans can have only a very slight effect on local climate (e.g. from land clearing or heat generated in cities). What is contested is whether increasing atmospheric carbon dioxide will have a catastrophic effect on global climate and whether the science is settled in the new field of climate science.

The last chapter of this book gives one hundred and one questions that young people should ask their teachers to ascertain whether they are being taught to think rationally, critically and analytically. If pupils are being fed activist propaganda, then they increase their chances of being unemployed and unemployable. Pupils can find out very quickly if their teacher is a dope or if the teacher is feeding them with environmental activist propaganda. This book is also for parents to ascertain whether their children are being fed environmental advocacy and political propaganda rather than being taught to think. They should ask children's teachers some of the questions.

This book is also for the punter who has a good nose for political spin and nonsense. The punter is blessed with common sense and the punter will be paying for the economic fallout from the emissions trading and carbon tax funding scams so strongly supported by banks. The punter should ask some of the questions on talkback radio when some clown is claiming that we are all going to fry-and-die. The punter should ask some of these questions in *Letters to the Editor* in newspapers to demonstrate the weakness of the human-induced global warming fad. The punter should ask some of these questions at a public meeting and should ask questions of their politicians. Don't believe me, ask a few questions and you will very quickly work out that we are being fed total nonsense about human-induced global warming.

If you are a school pupil, have you ever thought that some of what you are being told is incorrect? Have you ever thought that your teacher might be using the captive audience for political and environmental activism? Have you ever thought that what you are being taught is totally useless for life? Here is your chance to check out such thoughts. Maybe your teacher knows nothing about the planet and yet is teaching you about human-induced global warming. A teacher might answer your question with "I don't know". This is fair enough and hopefully that teacher would do a bit of work and come back the next day with an answer. There are simple innocent questions that you can ask your teachers. Before you ask such questions, you must know the background and know the answer to your own question.

If your teacher waffles on about consensus, settled science, the Intergovernmental Panel on Climate Change (IPCC) and scientific academies, tipping points, carbon pollution, human-induced global warming, sea level rise, irreversible global warming, moral issues, ocean acidification and future catastrophes, then you know you are being conned. Natural phenomena are not determined by the votes of human beings, regardless of their formal qualifications. In my experience as a school teacher and later with more than four decades as a university lecturer, I have always welcomed difficult questions. These come from the best students. The asking of questions could easily get you expelled from school in today's world, especially if you expose your teacher for being a political and environmental activist. You might point out to your teacher that surveys show that there is a consensus about a "Carbon Tax". People don't want it. Why do politicians think a consensus in science is important yet don't seem to think a consensus in the community is important?

Parents can ask the same questions of teachers to find out whether their children are getting a solid education or being fed propaganda. There are many people in the community that "believe" in human-induced global warming. Let's just forget for the moment that the word "believe" is a word of religion and politics, why not ask a few simple questions to find out whether this "belief" is based on knowledge or whether the believer has been conned by the relentless propaganda campaign by the government, the Green Party, advocates, political activists and those who choose not to think. One wonders if Green Party supporters have ever met the average person because they behave like old-fashioned wowsers wagging their fingers at us.

The punter is not stupid. Yet our politicians, bureaucrats, scientific activists, environmentalists and the climate commissioner have all treated the punter as a fool. The ABC, BBC and major newspapers have never ceased to tell us that we are going to fry-and-die if we don't change our ways. They have deliberately given no oxygen to the large group of scientists who see little or no evidence for human-induced global warming. The internet has enabled inquisitive punters to see that there is an army of eminent scientific dissenters out there and that the evidence for human-induced global warming is very weak. Only a few radio commentators have investigated the issue and have concluded that those promoting the human-induced global warming story are not telling the truth. The warmist arguments have been shown to be full of holes and the listening audience has quite rightfully wondered why there is so much fuss when there is so little evidence. The response in Australia from government, environmentalists and competing media has been to call for a media investigation in order to silence differing opinions. No wonder some regard the issue of climate as one of freedom.

The Australian government's propaganda-advertising program to persuade the punter about the merits of a carbon tax is deceitful. And the punter knows it. One such advertisement used as a backdrop an English power station that has been shut for half a century. The advertising implies that coal-fired power stations pump black soot into the air. They don't. Footage of coal-fired power stations show condensing steam and not the tall stacks that emit colourless odourless plant food. The government demands that we "respect the science" yet at the same time they mislead and deceive and give us no respect.

The punter's natural scepticism is reinforced by a major survey in the US. The Rasmussen Poll shows that a large majority of Americans think that human-induced global warming has not been proven and that warmist scientists may falsify information trying to prove their theory is correct. Since the Climategate fraud was revealed in late 2009, belief in human-induced global warming has dropped by 10%. Americans are very much aware that warmists exaggerate and that the doom-and-gloom picture is not the reality they experience. Through

US eyes, constant claims that there is a consensus in science and that the majority of scientists support the warmist view has worn thin and the public does not buy this any more. As a result, the public asks what other parts of the human-induced global warming mantra is fabricated. The warmists' stretched their credibility even more by the changing of the words "global warming" to "climate change". The US punters are awake to the fact that they are being conned. The environmental activists' standard procedure of attacking the man rather than answering simple questions of science has not impressed those surveyed. They want to hear the arguments and wonder what the climate industry is trying to hide. In short, the public has woken up to the hypocritical unctuous sanctimony and self-righteousness of high-profile environmentalists.

The whole basis for human-induced global warming can be found in just one chapter (Chapter 9) of the IPCC Report AR4. This chapter states that we dreadful humans are responsible for global warming and that a terrible fate awaits us. How do we know this? From a computer! But wait, it gets better. The computer models, codes and data for these frightening predictions are secret. Even Australia's CSIRO and Bureau of Meteorology also use secret computer codes.

If these predictions are to be taken seriously by governments, scientists, the public and the media, then the computer models, codes and data of computer models should be freely available to enable validation. This is what happens in the commercial world in the case of due diligence studies. To make matters even more bizarre, that once great organization the CSIRO now has legal disclaimers protecting them if people make decisions based on their scary predictions and the alarmism turns out to be incorrect:

> This report relates to climate change scenarios based on comput-
> er modelling. Models involve simplifications of the real processes
> that are not fully understood.
>
> Accordingly, no responsibility is accepted by the CSIRO or

the QLD government for the accuracy of forecasts or predictions inferred from this report or for any person's interpretations,
deductions, conclusions or actions in reliance on this report.

Any scientific document that needs a legal disclaimer is clearly
non-science and should be treated as nonsense. To use words such
as scenarios, forecasts and predictions is misleading. The CSIRO is
admitting that their climate work is meaningless. If governments are
to make multi-billion dollar decisions and change whole economies
based on these scary predictions, then why can't the computer models,
codes and data be checked as part of a transparent due diligence
examination? If a private corporation made a similar projection or
prediction and was not transparent, it would be considered fraud. Until
the IPCC computer models, their codes and their data are available
for all to see, then the whole complex edifice espoused by the IPCC
can only be fraud. To make matters worse, most of the temperature
measurements used in the models have been "adjusted" before they
are massaged by computers. Why haven't the media asked the obvious
questions?

Only one media person has chased down the question of
measurements in the IPCC Report AR4 Chapter 9. It was Alan Jones
(Radio 2GB, Sydney; 25th May 2011) in an interview with David
Karoly.

> *Jones:* Is there any empirical evidence proving human production
> of carbon dioxide – as distinct from nature's production – caused
> global warming? Is there? In these reports? Yes or no?
> *Karoly:* Yes.
> *Jones:* Now where would I find that in Chapter 9, that's your
> chapter.
> *Karoly:* Sure. You would find that evidence in the peer-reviewed
> scientific studies and in the data…
> *Jones:* But where in Chapter 9?
> *Karoly:* So…
> *Jones:* Where in Chapter 9? Where can I open Chapter 9, because
> I looked at it, where if I open Chapter 9 is that evidence? Where
> is it?

Karoly: It's… I can't tell you the page number because I don't…

Jones: No, no. It's not there. It's not there.

Karoly: What. No, Alan.

Jones: It's not there. You, the Chapter review editor. It's not there. That's why you can't tell me the page number. The evidence is not there.

Karoly: That's not true, Alan.

Jones: Well I've got scientists on stand-by who are going to listen to all of this so your reputation's on the line when you say that. I'm telling you, Chapter 9 is your Chapter. You in fact were the Chapter's review editor and you can't tell me where the evidence is.

Karoly: Yeah, I can. Would you like me to tell you where the evidence is? The evidence is in the spatial patterns and the time variations of temperature changes in the observations…

Jones: Whoa, whoa, whoa. Chapter 9, Chapter 9, David, is the Chapter. It was originally Chapter 12 in the 2001 report. In the 2007 report you were the review editor of this chapter on the direction… on the detection of climate change. It's now called "Understanding and attributing climate change". Now, to understand climate change you need to know what evidence there was for all of this. In Chapter 9 it's not there.

Karoly: No Alan, it is there. So would you like me to tell you which figure in particular in Chapter 9 shows that evidence? It looks at the pattern of climate variation over the last 50 and the last 100 years and what it does is it makes an evaluation or assessment. It talks about how climate has changed compared with what we'd expect from greenhouse gas variations, it also looks at other factors. Factors like changes in sunlight from the sun, changes in the effect of volcanoes, natural variations like El Niños, natural variations like the Pacific Decadal Oscillation and what it shows, what it clearly shows, is that the patterns of change are outside natural variability, aren't due to changes in sunlight from the sun and we can see that sunlight from the sun would cause more warming in the daytime when the sun's really important but we've actually observed more warming at night. We've seen changes in the temperatures in the lower atmosphere and the upper atmosphere which clearly show that the changes are due to the increases in greenhouse gases and aren't due to natural variability and aren't due to other factors. And we've…

9

Jones: Chapter 9, Chapter 9 doesn't contain any of that detail. Can I go on?
Karoly: Yes, it does.

No wonder Alan Jones gets pilloried by the loopy left. He does his homework and asks the important questions. No wonder he has the biggest listening audience in Australia. Karoly gives the game away, despite waffling on about science, by talking down to Jones and not answering the question. There is no empirical evidence in the key IPCC chapter to show that human emissions of carbon dioxide drive global warming. You heard it from the review editor for the IPCC's Chapter 9. Game over. We are being conned.

The IPCC assumes that the Earth will amplify the slight warming from an increase in carbon dioxide. They also assume that the climate system is unstable, which would be the first time in 4,500 million years of Earth history. The IPCC predicts that amplified warming will occur from an increase in water vapour over the tropics. Water vapour is the main greenhouse gas in the atmosphere. These predictions are not supported by measurements and are nothing more than speculative assumptions. Some scientists suggest that there will be far less warming than the IPCC predicts because they consider that the climate system has been changed beyond natural variability by human activity. The focus on the IPCC has been to show that humans change climate, not to understand climate. The effects of clouds, natural oscillations in the Earth's climate system and solar-cosmic influences are poorly known. The IPCC makes long-term predictions based on "adjusted" measurements over a very short time period that ignores natural cycles of warming and cooling. A reasonable person would conclude that speculative assumptions based on incomplete and adjusted short-term measurements are not the way that science or public policy should operate.

Planet Earth is extraordinarily complex. There are a large number of known factors that drive climate. Interactions of these factors can produce surprises, new knowledge is being discovered every day and there are many things we do not know. The science is not settled. Most carbon dioxide is emitted by nature. And always has been. Ever since we humans lit our first fire, we have been adding carbon dioxide to the atmosphere. It is because we add carbon dioxide to the atmosphere that we have food, water, transport, metals, plastics, concrete, heating, cooling and lighting. Without carbon dioxide emissions, we would have no modern life. Those trying to dictate that we emit less or no carbon dioxide do not lead by example. They do not live in caves as hunter-gatherers. This is to our advantage. Have a look at the leading lights of the climate industry and imagine them naked at a cave entrance. This is far scarier than the prospect of frying and dying from global warming. They flit around the world in carbon dioxide-emitting jets to attend climate conferences in exotic places. We've had Rio, Kyoto, Bali, Copenhagen, Cancún and now Durban. Why do they have jaunts? Why can't such conferences be held using the normal international electronic conference facilities or Skype?

Some of these people are telling us that as a result of our carbon dioxide emissions, sea level will rise tens of metres, yet these are the very same people who live at sea level in expensive houses bought from money earned from banks, governments and carbon dioxide trading houses during their global warming scare campaigns. The whole basis for the global warming propaganda is that there is an underlying assumption that the Earth was a pleasant benign place before industrialisation and that any change must be due to us. The climate industry claims that the last 100 years of climate change is unprecedented and just seem to conveniently ignore history that shows us periods of great warmings and coolings before industrialisation. This sort of deceit underpins the whole human-induced global warming movement.

There is actually no such thing as global climate. There are global climate zones. One classification has thirty such zones. Every part of

11

every country has a unique climate. We cannot compare changes in the weather and climate in the south of France, northern Alaska, the heart of Africa, the Great Lakes of USA, the tropics of Indonesia and the desert of Australia. Climate is defined as an average of weather over some 30 years but this too does not give variability as there are cycles that are far longer than 30 years.

Governments, activists and the ignorant are claiming that carbon is pollution. Carbon is element number six in the periodic table, it is natural and it forms the basis of all life. Carbon dioxide gas is natural, colourless, tasteless, odourless and non-poisonous. The air in any house or office normally has more than twice the atmospheric carbon dioxide content of the air outside. Our breath has more than one hundred times the carbon dioxide content of the air. Limestone and marble contain 44% carbon dioxide. About 97% of annual carbon dioxide emissions are natural and yet it is claimed that the 3% (i.e. human emissions) drives global warming. And even if it did, it has yet to be shown that a warming on Earth would be catastrophic. Carbon dioxide is plant food and without plants there is no life on Earth. We should be celebrating carbon dioxide, not demonising it. How much carbon dioxide in the atmosphere is too much? It is certainly not 0.02% at which most plant life dies. It may be the current level of less than 0.04%. It may be 0.1%, the amount in glasshouses used to accelerate plant growth and reduce water consumption. Whatever it is, we humans just cannot decide the optimum amount of carbon dioxide in the atmosphere and hope like hell that the atmosphere obeys our decision. If we really must demonise an element, leave carbon alone and pick on something totally useless such as technetium or promethium.

For pupils, parents and punters at a party, it is good to show people how much carbon dioxide they actually emit. Bubbles in wine, beer, soft drink and mineral water are carbon dioxide. Why not ask someone whether "carbon pollution" is dangerous and then point out the bubbles. To make the glass requires soda ash, sodium feldspar, dolomite, quartz and boron minerals. Crystal glass has lead oxide (produced by emitting carbon dioxide), bubbles are removed from glass with arsenic oxide (produced by processes that release carbon dioxide) and the whole process of exploration, mining, transportation, sizing, storage, conveying, weighing, mixing, feeding to a furnace, reactions to form glass, saving heat by recuperation or regeneration, shaping, annealing and finishing to produce some 800 different types of glass is not that easy. Every single step of the process involves the release of carbon dioxide. How can you possibly hold a glass in your hand and claim that carbon dioxide is pollution? If you break the glass, you are unlikely to recycle it because there are so many specific types of glass. So even more carbon dioxide is emitted to make new glass. If you do recycle the glass, then carbon dioxide will be emitted to produce the energy to re-melt the glass. Maybe you could look at the glass in a window. Or a glass-topped table. The glass is extremely flat because liquid glass was poured on molten tin. The exploration, mining, processing, smelting and transport of tin results in large emissions of carbon dioxide.

If you are somewhat uncouth and drink out of a can, then you probably don't care that the can is made of aluminium and some magnesium. A huge amount of energy is needed for the making of aluminium and most of this energy comes from the burning of coal. The can is stored energy and its manufacture emitted carbon dioxide into the atmosphere. Again, a huge amount of carbon dioxide is emitted in the exploration, mining and transport of bauxite. Conversion of bauxite to alumina normally uses energy from coal-fired power stations. The conversion of alumina to aluminium involves massive amounts of energy, generally coal-fired, and various fluorine-bearing energy-saving fluxes.

Maybe you will just stick to the prisoners' diet of bread and water to lead by example. Not that easy. Making of bread emits carbon dioxide into the air. The oven is powered by coal-fired electricity, gas or wood which, upon burning, release carbon dioxide to the air. For water, we need to construct dams, create metal pipes and pump water, and these processes again pour large amounts of carbon dioxide into the air. There may be music at the party. This requires energy. And on we go. There is no escape. Our life is far better now because of carbon dioxide.

Everything we use or do involves energy and the release of carbon dioxide. The cheapest energy derives from the burning of coal and it is total hypocrisy for a person in the Western World to live a Western life and yet tell others to live a more basic life. This is exactly what the Green Party does. Even in Earth Hour, burning candles emit carbon dioxide and coal-fired power stations still need to be kept running. It is most appropriate that the darkness is worshipped in Earth Hour and that there is a lack of recognition that humans have delivered the fundamentals of life to billions of people at a price they can afford by having carbon dioxide emitting energy. We should worship carbon dioxide on the altar of the modern world.

If Australia, Britain or other countries are to have a tax to reduce human emissions of carbon dioxide, then it needs first to be demonstrated that human emissions of carbon dioxide produce global warming. If they do indeed produce global warming, then either you adapt or you live a primitive hunter-gatherer life as all food, transport, heating, cooling and housing in today's world derive from coal-fired energy. I await the attempt by the great and glorious Department of Climate Change and Energy Efficiency to lead by example and have no central heating in Canberra's winter. I have yet to see a politician, environmental activist or banker give us their wisdom about the environment from the entrance to their cave rather than tell us how to live our life after a carbon dioxide-emitting business-class trip at our expense. Those scientists that preach doom and gloom do so from computer models using other people's original data. Furthermore, the methods unpinning the models (e.g. Mann's "hockey stick") have been shown to be mathematically flawed. They should get away from their computer and go outside and make measurements to validate the numbers they blindly enter into their computers.

As the planet has been warming for the last 330 years, I am at a loss to understand how increasing taxation, bloating the bureaucracy and creating the opportunity for opaque international carbon trading scams can save the planet from a natural process that has been happening for all of these 330 years. What are they trying to save the planet from? A warmer climate? History shows us that warm climates give us more prosperity and greater crop growth. We have the farce of the Federal Government wanting to impose a carbon tax on big polluters (i.e. mainly coal-fired power stations) yet in many places State governments generate electricity from coal. Will we see a Federal government taxing a State government for "carbon pollution"?

There is great pressure to generate ideological electricity from unsightly inefficient wind farms or solar stations that do not provide electricity when we want it. We gave up using wind generation 150 years ago. And why did we give it up? It was hopelessly inefficient and it is only being used today because of huge subsidies. Photovoltaics were invented in 1839. We have had more than 170 years to create large amounts of electricity by this process and have failed because solar power is hopelessly inefficient. If someone could invent far more efficient photovoltaics, they would end up the richest person on the planet. The motivation for solar power could not be better. However, we should abandon thinking about solar power and use tried and proven methods for generating the large amounts of electricity we need. Are the environmental activists aware that the energy used to create, maintain and retire wind and solar power stations is greater than the energy that they produce over a 25-year period? Sea breezes and sunbeams are not economic ways to generate the large amounts of electricity we use around the clock and they are probably more environmentally damaging than conventional electricity sources.

The amount of carbon dioxide in the air is increasing. There is a huge controversy in the scientific literature about where carbon dioxide comes from, how long it stays in the air and where it goes. Calculations by the IPCC can't find about half the carbon dioxide they estimate to be emitted to the atmosphere by humans. Maybe the estimates are seriously wrong, maybe carbon dioxide is used by life far quicker than the IPCC would have us believe or maybe much of the carbon dioxide comes from elsewhere. Maybe the missing carbon dioxide goes into oceans quicker than the IPCC would want to believe. Maybe, Heaven forbid, the IPCC just exaggerates. Matters are not helped by positioning the main carbon dioxide reference measuring station on an active volcano in Hawaii that emits carbon dioxide. The measuring station is at altitude and therefore does not measure carbon dioxide at sea level, assumes that there is mixing between the Northern and Southern Hemisphere atmospheres and does not measure carbon dioxide dissolved in water vapour. Furthermore, more than 80% of Hawaiian measurements are rejected and all of the pre-1960 non-Hawaiian measurements are rejected by the IPCC.

Historical measurements of carbon dioxide show great variations with peaks far higher than the current atmospheric carbon dioxide content. This spoils the fry-and-die story and so some 90,000 historical measurements of carbon dioxide before 1960 are just simply ignored by the climate industry because it is claimed that they are inaccurate. However, modern ice core measurements don't have this problem and differ substantially from land measurements.

If Australia reduces "carbon emissions" by 5% by 2020 or has a tax on carbon dioxide emissions, will it have an effect on planet Earth? This can be calculated. Humans emit 3% of annual total carbon dioxide emissions, the rest are natural and mainly from ocean degassing. Australia emits 1.5% of the total human emissions. If the atmospheric measurements of carbon dioxide are projected from

2011 (389 parts per million) to 2020 (412 parts per million), then a 5% emissions reduction by Australia will lower the atmospheric figure to 411.987 parts per million. This would lower global air temperature about 0.0007°C by 2050. This is nothing. Even if Australia stopped emitting carbon dioxide today, the decrease in global temperature would be 0.0145°C by 2050. Temperature is constantly changing by far greater amounts than this. In fact, if a Tasmanian moved from Hobart to Darwin, the average temperature rise the Tasmanian would encounter would be 18°C. If a Finn moved from Helsinki to Singapore, the average temperature rise the Finn would encounter would be 22°C. Variation between summer and winter temperatures in the outback is more than 30°C and in high latitude areas more than 50°C.

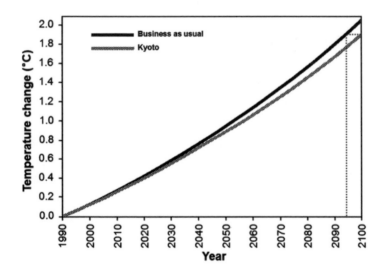

Figure 1: *The effect on temperature of 37 Western countries reducing emissions by 5% from their 1990 emissions according to the Kyoto Protocol. In effect, the 0.2°C temperature reduction ignores natural climate change and show the pointlessness of trying to legislate to change climate.*

17

Let's just ignore the annoying little fact that there has been no global warming since 1998 yet carbon dioxide emissions have been increasing. If every person on Earth was to try to stop the planet warming by the predicted 0.24°C from 2011 to 2020 as a result of human carbon dioxide emissions, it would cost every man, woman and child on the planet $60,000. This is 60% of global GDP and 22 times the maximum estimates for doing nothing about possible global warming. Are you prepared to pay $60,000 for less than a quarter of a degree temperature change? I don't think so. Especially as the planet has been cooling in the 21st century and there is no correlation between temperature changes and human emissions of carbon dioxide. Without correlation there can be no causation.

Now you can see why the climate alarmists, governments, industry and banks don't want their policies questioned because it could stop the flow of money from your pocket into theirs. The same people who brought us the Global Financial Crisis are salivating at the prospect of an even bigger and better scam. It is the young people reading this book who will pay. It will stop them owning a house or having the same standard of living as their parents. Government action on human-induced global warming has nothing to do with climate or the environment. It is a method to take money out of your pockets, to keep the big banks happy and to take away some of our basic freedoms. It is a game of power (and I don't mean electricity).

Calls for moves towards clean energy are driven by those who will profit from its adoption. Why on Earth should we pay for expensive clean energy (that is not clean) when we already have cheap clean energy? In Australia, the black-coal power stations are really marvellous. Burning coal heats water to make steam. This steam drives turbines that produce electricity. The boilers are 96% efficient, exhaust heat is captured and reheats water, the 4% heat loss is through the boiler wall, some moist air is condensed, precipitators remove 99.99% of small particles derived from coal burning and carbon dioxide escapes up the stack to the atmosphere. And what does the carbon dioxide emitted from the stacks do? Fertilises plants. What can be wrong with this? Plants do not know whether the carbon dioxide they use is from human emissions or from natural emissions. Carbon dioxide is carbon dioxide.

Coal-fired power stations generate electricity efficiently with some 8% of the generation cost being the cost of coal. Cheap electricity produces jobs. The world is envious of Australia's cheap electricity. China is shutting down small old dirty inefficient generators and building a new large modern efficient coal power station every week. Australia did that years ago. If Australia had a 5% decrease of carbon dioxide emissions, the increase in Chinese emissions alone is expected to be over 100 times as large as Australia's reduction hence, whatever Australia does, it will have no effect on the planet. Australia, like the USA, has coal-fired power stations because we have hundreds of years of coal reserves. It is crazy to think that what is sinful in Australia is good in China. The politics of the Green Party clearly are not pragmatic as they do not involve global solutions to a perceived global problem. But then again, history tells us that ideology and economics have never been good bedfellows.

Imagine a huge conference room filled with public servants in the Department of Climate Change and Energy Efficiency slaving away trying to save the world from the ravages of computer-predicted human-induced global warming. Can global climate be changed from a room in Canberra? No. This is surreal, pointless and a huge waste of money. This room would not be smoke filled, any use of robust language would result in counselling, participants would be drinking organic water and non-synthetic carpets would be tofu stain-resistant. Meetings would finish five minutes before lunch or knock-off time. The term smoko would be banned. Just think how many folk would have to take stress leave after being employed for years to change the planet's climate only to discover that nature does not obey Canberra's legislation. Australia and the UK are the only countries with a

Department of Climate Change. What on Earth do the hundreds of public servants in Australia's Department of Climate Change and Energy Efficiency do each day? They certainly don't change climate, they certainly cannot change climate but they certainly consume huge amounts of your money. How can these public servants come home at night and proudly feel that they have earned their money? Maybe they think that dreaming up new ways to control people is actually work. What really is the point of a nation-bankrupting carbon tax that only decreases global temperature by 0.0007°C by 2050? Has it really been thought through?

Australia exports hundreds of millions of tons of steaming and coking coal to other countries. Are these countries going to stop taking Australian coal because they want to save the planet? A bit of reality would go a long way. Can you imagine the consternation when bureaucrats finally learn that plants just don't care whether their carbon dioxide food derives from natural sources or from human emissions? This could lead to an emotional breakdown followed by months of paid stress leave. Can you imagine the disappointment that bureaucrats would suffer when they learn that no other country has imposed a bankrupting carbon tax that is all pain for no gain, that Australia has lost its natural competitive advantage and that tax-paying industries that relied on cheap energy have now closed down. There is now such an investment of time, emotional energy and our money by bureaucrats and scientific activists that they must try to have some policy that will actually work. The warming of 1977 to 1998 was politically exploited. Now that there has been no statistical measurable warming for over 15 years, activists, advocates, bureaucrats and politicians just can not abandon their treasured but invalidated theory of human-induced global warming.

Scientifically illiterate politicians face huge pressures to do something to save the planet. Instead of educating the public about the realities of climate past and present and seeking a diversity of opinions, politicians capitulated under pressure. We should be mindful that in 1910, Argentina was the wealthiest nation on Earth. To live in a lucky country is just not good enough. Argentina went out backwards through bad policy. Australia is a lucky country but bad decisions and poor policy could very quickly push Australia into poverty, as happened with Argentina.

Humans have adapted to live by the sea, in the mountains, in deserts, on ice sheets, in the tropics and in the artificial reality of cities. In warm times, history shows us that humans thrived. The biggest stresses humans endured from climate were the changes from interglacials to glaciation. During glaciation, the population decreased and food sources diminished. History shows that we have numerous extreme weather events during times of cooling, not warming. We are now in an interglacial facing the next inevitable glaciation. Humans have shown many times in the past that they can easily adapt to warming. We saw from the Minoan, Roman, Medieval and Modern Warmings that life is far better in warm times than in cold times. We now have technology that makes it easier to adapt than in former times. If planet Earth is warming, then why is warming such a problem and how can a tax make it go away?

The Great Barrier Reef is used as the poster pinup for human-induced global warming. It appears that we are wrecking havoc, the reef is bleaching and we had better hurry up to see the Reef before it is totally destroyed. By us. Coral reef expert Ove Hoegh-Guldberg told us in 1998 that the reef was under pressure from global warming and that the reef had turned white. In 1999, he warned that global warming would cause bleaching of the reef every two years from 2010. In 2006, he warned that global temperatures meant that: "between 30 and 40 per cent of coral on Queensland's Great Barrier Reef could die within a month."

He later stated that there had been a minimal amount of bleaching of the reef. He also stated that the reef had made a "*surprising recovery*".

The alleged recovery would only have been surprising for Hoegh-Guldberg if he believed his own exaggerations. The Australian Institute of Marine Science showed that in a study of 47 reefs over 1,300 kilometres of the Great Barrier Reef, that the coral cover was stable and there had been no net decline since 1995. The ABC was telling us in 2002 that 10% of the Reef has been lost to bleaching since 1998. In order to persuade voters to support a "Carbon Tax", Australia's Prime Minister tells voters that global warming is already killing the Great Barrier Reef. She does not tell us that it has survived warmer times in the past. Is the ABC operating as an environmental advocacy organisation in the absence of science? Why is Australia's Prime Minister, with access to a great diversity of scientific advice, not telling the truth?

The story of large long-lived corals is different from what we see on television. A 337-year record shows that there were wetter conditions and higher river flow into the reef in the late 17th to mid 18th and in the late 19th centuries. Drier conditions were in the late 18th to early 19th and mid 20th centuries. We are told that by emitting carbon dioxide, we humans are going to make the climate, wetter, drier, colder and warmer. Whatever the weather is, apparently it is our fault. The Great Barrier Reef shows that it was extremely resilient to wet and dry times since well before human industry was emitting carbon dioxide.

In Australia, the major natural hazards have been droughts, floods, bushfires, storms and cyclones. Hundreds of people have been killed in cyclones, floods and bushfires over the last 200 years. We have had one fatal earthquake and can expect more. There have been deaths from landslides. The last volcanic eruption was 500 years ago and every few hundred years Australia is hit by a tsunami. Like all other countries, we have more premature deaths from cold weather than from warm weather. The US National Weather Service has kept statistics on weather-related deaths since 1940. The annual number of deaths from tornadoes, floods, hurricanes etc varies. For example, the number of people in the US killed by extreme weather events in 1972 was 703. In 1988, it was 72. There is a clear long-term trend: the number of weather-related deaths has dropped dramatically despite the fact that the US population has more than doubled since 1940. Studies on floods show that intense floods are less common in recent years and that flooding cannot be related to global warming. Increased flood costs may well be because we have become wealthier and more expensive buildings are now built on flood plains.

Agriculture and forestry turn atmospheric carbon dioxide into carbon compounds. Do those in the climate industry eat food or use anything made of wood? Food is carbon-based and, as a result of metabolising our carbon-based food, we exhale carbon dioxide. Considering that Australia exports food for about 60 million people and a "Carbon Tax" will reduce food production, ask your friendly bureaucrat which countries they have decided to starve. They have already had a good attempt at this by stopping meat supplies to our close neighbour. How much carbon dioxide and complex carbon compounds can bushfires put into the atmosphere. One big bushfire can have a profound effect on the carbon budget.

Australia produces some 1.5% of the world's human emissions of carbon dioxide. China and USA account for about 40%. If Australia reduces emissions by 5%, it would be cancelled out by a 0.3% increase in China's emissions. Whatever Australia does will have absolutely no effect whatsoever on the planet. Get used to it. Considering that Australia exports metals, it is no wonder that we have high emissions per capita because smelting releases large amounts of carbon dioxide to the atmosphere. We produce these metals for the global markets because we are a stable country, have a stable energy supply and have cheap sources of energy. A "Carbon Tax" will change all of this.

INTRODUCTION

This book is a sequel to *Heaven and Earth: Global warming – the missing science* (Connor Court, 2009). In that book I showed that our planet changes all the time, it is very resilient to anything thrown at it and that past climates have changed by the interaction of changes in the Sun, the Earth's orbit, stars and major Earth processes. Changes in the oceans, ice sheets, atmosphere and life just don't happen by themselves and are driven by forces far larger than anything we humans can muster. The key point is that climate has always changed, that we cannot even hope to understand modern climate without trying to understand the past (i.e. the missing science).

If the science of human-induced global warming is so strong, then why is it necessary for the climate industry to engage in fraud, exaggeration, obfuscation, personal attacks, spin and the demonising of dissent? Why do the climate industry scientists need to change field measurements before these are fed into computers? Why were so many measurements of carbon dioxide, temperature, sea level and weather patterns ignored or rejected? Why is it that the climate industry wants to ignore the Minoan, Roman and Medieval Warmings when it was warmer than now for hundreds of years, when sea level did not suddenly rise, when people did not die from global warming and when environments were not devastated? There was no coal-burning industry then to make the planet warmer. Why is it that the climate industry ignores the long history of planet Earth during which time climate continually changed, the six major ice ages started when the atmospheric carbon dioxide was higher than now and sea level rose and fell by a mere 1,500 metres?

Since *Heaven and Earth* appeared, there have been international climate conference love-ins at Copenhagen, Cancún and Durban.

Tens of thousands of participants jetted in, stayed in expensive hotels, preached doom and gloom and unsuccessfully tried to organise economic suicide pacts. It was so comforting to have folk sacrifice themselves to go to an expensive resort like Cancún and try to work out ways to take money from us. According to the UK Taxpayers' Association, Copenhagen cost as much as the GDP of a small African nation. The Australian taxpayer funded more than 100 bureaucrats to attend the Copenhagen group-think. And what was decided? Nothing. What did we learn? Nothing. This is not a new phenomenon. In *Extraordinary Popular Delusions and the Madness of Crowds* (Charles MacKay, 1841), we read:

> Men, it has been well said, think in herds; it will be seen that they go mad
> in herds, while they only recover their senses slowly, and one by one.

I suspect that we are in one of those periods when people are recovering their senses one by one after a period of herd madness.

Heaven and Earth was a best seller. I predicted that I was going to enjoy irrational attacks and I did. Imagine my trauma when activists savaged me with the accusation that the book was not peer-reviewed. Which alarmist book of Tim Flannery has been peer reviewed? Had any of my critics ever written a book? No. Do those making the accusations understand how the peer review process operates? Have they been editors of, or submitted papers to, major scientific journals? Those making such an accusation demonstrated that they had not read the book because the reviewers were listed and thanked. Like many subjects on blog sites, these accusations were repeated *ad nauseam* without checking the primary source. Trust me, being savaged with a feather is not for the faint hearted. What all critics just happened to fail to mention was that either they or their institution had a vested interest in the human-induced global warming scam. If they acted like this in the commercial world, they would be behind bars.

Please, dear reader, continue to share my suffering. One Michael Ashley attacked me in a major national newspaper. His field of expertise is "electronics, optics and computing (both computer hardware and software) with the goal of building new and interesting astronomical instruments" yet he criticised me on geological matters. Furthermore, he appears to have no expertise in climate. Just ask my publican, also a geologist. I was completely devastated that a modern-day clock maker promoted my book via criticism. I just could not leave the front bar for weeks.[2] UK-based George Monbiot, catastrophist journalist, irrationalist and unintentional comic may have once had a little knowledge about zoology but chose to attack me on carbon dioxide emissions from volcanoes. And his source: one scientific paper. Only one and not even written by him. Why bother to read the hundreds of other scientific papers because they don't fit a pre-determined ideology? Why bother to get an understanding of a complex subject when one can read just one paper that supports his ideological view? And what did this paper do? It looked at measurements from about 20 volcanoes and ignored more than 3 million volcanoes. Monbiot has since had to recant on most of his catastrophic predictions and admit that he was wrong on so many things that he hyped in the press.

Imagine how catatonic I became when an eminently unknown mathematician Ian Enting attacked my geology in a creationist-styled diatribe replete with numerous contradictions, a lack of critical thinking and fundamentalist ideology. His 15 seconds of fame was not well used. It was really scary. Have a look at his profile.[3] Terrified and almost destroyed, it took me some time to pull myself up from the floor to reach the front bar for some more revival juice. It was not easy.

And then, just when all had quietened down, another attack came in from left field. The Australian Government was trying to introduce a "Carbon Tax" so it gave its stooges a free kick at a Senate Estimates Committee. Can you imagine my bliss when the Bureau

---

2 Junction Hotel, corner of Silver and Argent Streets, Broken Hill.
3 http://www.ms.unimelb.edu.au/~enting/.

of Meteorology chief Greg Ayers claimed that *Heaven and Earth* had influenced Cardinal Pell. I was so overjoyed with the claim that one who has just been a mere Professor of Geology for more than 25 years could actually influence a Cardinal. I had always thought that this was the role of the Pope and the Almighty. With a spring in my step, I went back to my local to celebrate my change of fortune.

All critics ignore the fact that there have been past times well before industrialisation when it was far warmer than now. These warmer times could not possibly have derived from human emissions of carbon dioxide and these warm times were neither catastrophic nor irreversible. The critics have trouble with these bits and, rather than engage in discussion, they just totally ignore anything that does not fit their dogma. This is why the past is removed from all arguments and why the climate industry wants to remove history, archaeology and geology from their mantra. They claim that the very complex climate system is only driven by human emissions of carbon dioxide and not the 97% of natural emissions. Unquestioningly embracing the faulty and fraudulent IPCC ideology and then crushing any opposition, as was tried in the Senate Estimates Committee, has dangerous historical precedents. The use of Senate Estimates by bureaucrats to denounce those that don't embrace the official government doctrine shows the lengths that governments will go to ensure that science in Australia is in the services of politics.

The standard discredited arguments were used to attack *Heaven and Earth*. The IPCC was quoted as the authority. Argument by authority was shown to be fallacious 2,500 years ago and a report by a committee paid to find a given result is only proof of a lack of independence. The IPCC actually conducts no research, their summaries are prepared by activists, bureaucrats and politicians and the body of the report is prepared by scientists, activists and political advocates. These folk can hardly be viewed as independent. My view is even supported by those in the IPCC camp. For example, Hans von

Storch of the Meteorological Institute of the University of Hamburg has warned that the IPCC's warming story is far from independent and does not cover the spectrum of published information:

> The IPCC has failed to ensure that assessment reports, which shall review the existing public knowledge and knowledge claims, should have been prepared by scientists not significantly involved in the research themselves. Instead, the IPCC has asked scientists like Professor Mann to review his own work. This does not represent an independent review.

The IPCC Summaries for Policymakers are released before the body of the report. If journalists were inquisitive and had some scientific knowledge, they could easily show that the Summary for Policymakers is not a summary at all but states pre-ordained conclusions. The scientific body of the report contains contrary information. What journalists did not do was to read the 987-page Working Group 1 IPCC 2007 report. If they did, they would have found that the words "uncertain" and "uncertainty" appeared more than 1,300 times and that there are 54 "key uncertainties" that acknowledge limits to prediction of climate. The IPCC itself shows that the science is not settled, that there is no consensus and that little is known about the controls on the climate system. But all this is hidden in the small print and journalists have just not bothered to read what might contradict their own opinions. The recent amended statement from the Royal Society also expressed a key uncertainty by stating that they don't know how much warming could result from rising carbon dioxide levels. When it comes to climate, uncertainty rules.

Those scientists who may be independent or provide balance are excluded from the club of true global warming believers and hence the report does not represent the view of the overall body of science. When the IPCC releases a new report with great fanfare, legions of environmentalists, advocates in government offices and the unscientific uninquisitive media shout the latest catastrophic warnings. An IPCC report is not evidence, whether it is based on selected peer-reviewed papers or not. Furthermore, just because a scientific paper is peer-reviewed does not mean it is correct. The peer-reviewed scientific literature is full of papers that contradict each other so they can't all be correct. Most peer-reviewed papers become outdated quickly because of new information or because the phenomenon cannot be replicated. Peer review does not stop bad science being published. Scientific theories live or die on evidence, not whether or not they were published in the peer-reviewed literature. Some of the greatest elegant ideas that have stood the tests of time, replication and attempts to falsify were not peer-reviewed. The best 20th century examples are the works of Wegener, Einstein, and Watson and Crick.

Many of the critical references used by the IPCC were advocacy documents by pressure groups. These were certainly not peer-reviewed. Despite the massive volume of the IPCC reports, they do not show that carbon dioxide causes significant warming. My critics often called on consensus but it only takes one piece of evidence to show that a consensus is wrong. This is done all the time in the evolution of scientific thinking. A recent example was the discovery that a bacterium causes ulcers and the popular consensus theory on ulcers had to be thrown out. Science is based on evidence. Nature does not obey academies of sciences, bureaucrats, politicians or environmental activists.

The favourite political ploy used in attempts to silence me was to claim that the debate is over. There has never been a debate on human-induced global warming and the only reason there are attempts to stop debate is because the arguments in support of human-induced global warming have all been shown to be weak, concocted or discredited. The average punter can smell when something is on the nose. Find me a taxi driver who argues that human emissions of carbon dioxide drive climate change and, in order to stop dangerous global warming, we need a new tax. The precautionary principle was raised by critics but they can never answer how much should be spent to fix something that is not a problem. The same people who invoke the precautionary principle are great supporters of "green energy" yet they do not use the precautionary principle to argue for the shutting down of every wind generator because of their possible dangerous health and environmental effects. If, as the evidence shows, human emissions of carbon dioxide do not drive global warming, then we await an explanation as to why we should pay for carbon dioxide emissions, subsidise uneconomic sea breeze and sunbeam electricity and pay a "Carbon Tax."

Many climate industry critics claimed that I am not a "climate scientist" without pointing out that "climate science" is a synthesis of the natural sciences, not a science itself. Such critics didn't point out that geology is the only way to establish the history of climate on Earth. Climate just did not suddenly appear on Earth because we are alive. The modern climate is a result of past climates. All 19th and 20th century geology textbooks have large sections on climate. Geology is the ultimate climate science and to try to exclude geology from any discussion of climate shows that the climate industry advocates are misleading and deceptive by wanting to paint a narrow and modern history of the planet. The field of so-called "climate science" is a very recent creation by those who exclude history and geology, who exclude those with contrary ideas from grant-gobbling activities and who exclude nature from their activist conclusions. "Climate science" is not mentioned in 19th and 20th century textbooks and starts to make the occasional cameo appearance in late 20th century texts.

Each year dozens of cores are drilled for scientific purposes and millions of holes are drilled by the exploration, oil, coal and mining industries. It is the geologists in these industries who possess the largest body of data on past climates and are also the most vocal opponents of human-induced climate change. Why? Because they are practical practising scientists whose lives do not depend upon government research grants. Their lives depend upon field observations, empirical testing, measurements and successful results. Drill holes very quickly show how hopeless models are because a drill hole is a type of three-dimensional lie detector.

Oil geology depends on knowing what sea level and climate were doing at the time of sediment deposition. Some 40 years ago an Esso research geologist created graphical sea level curves (Vail Curves). These are still used as the benchmark for sea level changes. Such graphs are a window into past climate and give a sober view of modern climate. Coal deposits form close to the shoreline in cold climates and drill holes are used to determine the migration of the sea backwards and forwards across the land. Minerals geologists also determine ancient climate and sea levels, especially those exploring for heavy mineral sand deposits that formed in storms at maximum sea levels. Some of these mineral sand deposits occur hundreds of kilometres inland and over a hundred metres above the current sea level and hence knowledge of climate, sea levels and land level changes is required for successful exploration.

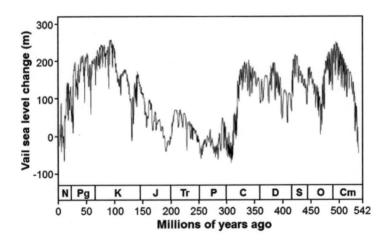

Figure 2: *Vail sea level graph showing massive fluctuations in sea level over the last 500 million years. Current sea level on this graph is 0. For most of the last 500 million years, sea level has been considerably higher than now, past sea level changes have been rapid and greater than the worst case computer predictions.*

Many other geologists chase rocks that have naturally sequestered carbon dioxide, as these too can contain valuable resources that we all use. Geologists are the only scientists who have a practical use for past climates and sea levels. They must get it right and the only way to get it right is to understand climate and sea level changes. If they get it wrong, they are out of a job because there are no research grants to keep them alive. If those in the climate industry get it wrong, life goes on and the research grants keep rolling in.

The story of the history of climate changes is preserved in marine and lake sediments, ice sheets and all rocks. For example, cores drilled through the ice sheets yield a record of polar temperatures and atmospheric composition ranging back to 120,000 years in Greenland and 800,000 years in Antarctica. Oceanic sediments preserve a record reaching back tens of millions of years and sedimentary rocks preserve records of billions of years of the Earth's climate history. Fossils of corals, coral and algal reefs and plants all tell stories about past climate, as do stalagmites, tree rings, land shapes and soils. Throughout time, carbon dioxide has been emitted by natural processes and almost all of the world's carbon dioxide that was ever in the atmosphere is now sequestered into rocks.

The leader of the warmist brigade is the IPCC chairman, Rajendra Pachauri. He is a railway engineer and author of soft pornography, and hence is eminently qualified to talk about climate. Those involved in frightening us witless about climate change in Australia can also hardly claim to be "climate scientists". Australia's climate commissioner has expertise in Papua New Guinean tree kangaroos, Ross Garnaut is an economist, Barry Brook has expertise in macrofauna extinction and Will Steffan is a chemical engineer. I guess Al Gore has expertise in something, although he managed to earn the minimum examination grade in Natural Science at university. The list goes on. Others who have no idea of how the world works (Andy Pitman) tried to reinvent himself as an "earth systems scientist", whatever that might be. Various mathematicians, who need to be removed with an angle grinder from their computers, suddenly became experts on everything (e.g. Ian Enting, David Karoly, Andy Pitman) yet don't seem to base their conclusions on data they have collected themselves (with all the sobering limitations that brings). Giving contrarians labels does not change the scientific facts. There is no book by a warmist clearly stating their scientific arguments for all to see. Why not? Why is there no warmist equivalent to *Heaven and Earth*? They claim that the "science is settled", that "debate is over" and that there is a "consensus" but we have yet to see their simple proofs put into a book for all the public to see. If such a book ever appears, it should be called the *Carbonist Moneyfest*.

As for politicians claiming that there is a "consensus", that the "science is settled" and that the "debate is over", it is probably wiser if these politicians stick to politics and not science. And as for bankers, investment houses and large corporations claiming to believe that humans change climate and that emissions trading systems should be established, just follow the money. Money tells the story. The climate industry, like any other, responds to incentives and there is a huge amount of money sloshing around for those who play the game.

And as for journalists operating as political advocates. What's new? Almost all journalists are ignorant of science. Asking a series of questions, commonly compiled by others, hides their ignorance. Questions normally are intertwined with moralising. This is exacerbated by the lack of clarity of thought demonstrated by sentence construction, the asking of questions with preordained conclusions and the exposing of the journalist's ideology during the interview. In 2007, Robyn Williams of the ABC *Science Show* when interviewed by fellow journalist Andrew Bolt claimed that a sea level rise of 100 metres in the next century was possible. The most catastrophic sea level rise speculation by the IPCC is 59 centimetres over the next 100 years. This was a massive exaggeration or concoction *ex nihilo* by

Williams and shows that the host of the ABC *Science Show* is out of contact with science and uses his program for unsubstantiated activist propaganda. If Williams was correct, then we should have already had a 5-metre sea level rise since 2007.

However, there was one serious technical error in *Heaven and Earth* (and this admission, just quietly, is only between the two of us). It was discovered by a former student, Peter Black OAM. He was Mayor of Broken Hill for 19 years, Member of Parliament for 8 years and a knowledgeable mineral collector. He gleefully pointed out that the chemical formula for plimerite I cited in *Heaven and Earth* was wrong. And it was. Why did no climate industry critic find this obvious error on the first page of the book? It was because they did not check facts and criticised the book on the basis of ideology. This is just one of the numerous examples to show that critics of *Heaven and Earth* had not even read the book.

Another former student (Tim Casey) was somewhat annoyed that I had forgotten to reference a key paper that shows there are millions of submarine volcanoes and these types emit huge quantities of carbon dioxide. There were hundreds of letters, phone calls, faxes and emails about *Heaven and Earth* from people all over the world who had read the book from cover to cover. Most of the letters were from the average punter. One can count on a saw miller's hand the number of letters that were not supportive. I guess angry people cannot write letters. The theme of most letters was: thank you, I always felt that human-induced global warming was nonsense and keep fighting the good fight. *Heaven and Earth* led to a busy national and international speaking schedule that prevented airlines going broke and put a huge amount of plant food into the atmosphere. Many correspondents expressed concern that human emissions of carbon dioxide are related to the increase in population and the increase in the standard of living. While this is true, emissions of carbon dioxide allow more food to be grown with the addition of less solid fertiliser.

It was interesting to note that critics (especially on blog sites) struggled with spelling, English expression, punctuation and good taste. Blogs perpetually repeated errors demonstrating plagiarism and a lack of originality of thought. Most critics were anonymous. Critical and analytical skills and clarity of thought were in short supply. There was no sense of humour displayed. This I can understand, as it is a pretty onerous job to be the self-appointed saviour of planet Earth. No time for humour when the catastrophic clock chime cometh. Scientific critics were few and far between, commonly had an undeclared self-interest, criticised technical areas outside their area of training and were very bitter. I guess these critics have invested so much emotional energy into a career of frightening the community witless that rationality becomes a perceived threat. Supporters were normally middle-aged, many were retired scientists and engineers, they wrote better English, were polite and displayed a better and broader education than the critics. It was interesting that the older people were concerned about the environment and economy that their grandchildren would inherit yet younger correspondents showed little concern about the future, wrote about the present and gave me the benefit of their experience and ideology. Many supporters expressed concern that children were not being taught to think and that the schools were being used for environmental activism.

So, why write another book and expose myself to the anger of self-appointed saviours, journalistic advocates, environmental activists, critics with a self interest and an army of angry people? It is because many correspondents asked me to rewrite *Heaven and Earth* for people under 20 years old and to underpin this new book with the science in *Heaven and Earth*. And of course, dear reader, this I do just for you.

1

HUMAN-INDUCED GLOBAL WARMING: WHY I AM SCEPTICAL

In the morass of Western political group-think on human-induced global warming, a prominent light of rationality shines through. It is President Václav Klaus of the Czech Republic. His voice in the wilderness has questioned the economics of dubious measures that, at best, would decrease possible future global temperature by a fraction of a degree but at a huge cost. He has argued that human-induced global warming is a new irrational ideological religion dictated to humanity, commonly by unelected people, with the accompanying loss of fundamental freedoms. Advocates of human-induced global warming live mostly in wealthy Western countries and have never experienced the horrors of war, genocide, starvation, deprivation, incessant propaganda and loss of basic freedoms. All we have known in Western democracies is freedom, which we take for granted.

The issue of human-induced global warming is about power and has little to do with the environment, saving the planet, creating a better world and freedom of speech. Except when travelling in communist and other totalitarian countries, all my life I have enjoyed freedom of speech. The present state of public debate on climate is such that the government-approved beliefs are virtually compulsory. Those imposing their apocalyptic doctrinal views upon us want no rational civilised argument (e.g. "the science is settled"), claim that there is a "consensus", attempt to denigrate, and vilify and marginalise those who question the dogma (e.g. "climate deniers").

I have been in science long enough to know that when there is a consensus from a scientific group, then they have a vested interest. This book is a direct challenge to those seeking power and control over us by using apocalyptic scenarios of the future climate. Advocates of climate change despise man's great achievements. What next? Book burning? (By the way, this book is printed on highly flammable paper to enable efficient book burning in the interests of the environment). If human-induced global warming is so certain, then any additional funding for research on climate change should be stopped, all climate institutes should be closed down and the billions of dollars of funds should be redirected to feeding a population that will be using reduced amounts of fossil fuels. I am sceptical of the climate industry pushing the bandwagon of human-induced global warming so hard when it is clear that institutions and individuals receive each year millions of dollars of taxpayers' money to sing from their master's hymn sheet. These individuals are my most bitter critics and have everything to gain by silencing contrarian scientists.

There is no better person to write the Foreword than President Klaus, a person who has lived through fascism, communism and now capitalism. President Klaus argues that we are now seeing the evolution of a darker side of capitalism that has some of the hallmarks of the communism which he endured for so long. For one who was born on 19th June 1941 during occupation and grew up in the Czechoslovak Socialist Republic, President Klaus has had ample experience of having thinking processes dictated to him. Because of his economic views, which transpired to be correct, he was marginalised in communist Czechoslovakia and lost his job. He was later employed at computer modelling, a process he knows first hand and treats with scorn. Dr Klaus entered politics after an illustrious career in economics and served two terms as Prime Minister and now two terms as President of the Czech Republic. Under his leadership, his country avoided the global financial crisis because of his policies and has avoided national bankruptcy from environmental scams. He brought the Czechs out from under communism and into a healthy capitalism.

President Klaus has lived through history (some of which he made), has seen the result of disastrous policies and is eminently qualified to argue that if we ignore history, then we do so at our peril. Communism taught him not to believe what he reads in newspapers and that the Western media must shoulder much of the blame for the popularity of the human-induced global warming doctrine. Dr Klaus saw that, under communism, the worst environmental disasters were created in Eastern Europe in the name of the people. He has been a Eurosceptic and was sceptical about the future of the Euro and kept his country's currency. For this he was again marginalised and again time has shown that he was correct. For his views on human-induced climate change, he has again been marginalised. This is why I asked Dr Klaus to write the Foreword of this book. President Klaus and six reviewers have read this book, their comments are gratefully acknowledged. The drafting was done under great pressure and with numerous changes by Rob Schroder. He is thanked for his professionalism and tolerance. Although this book is the sequel to *Heaven and Earth*, numerous public lectures, debates, radio programs and letters guided me to the questions that troubled the target audience.

I am sceptical about human-induced climate change because the empirical evidence from the history of planet Earth shows that natural climate changes have been rapid, large and unrelated to human activity. I do not accept the scary scenarios painted by the scientific proponents of human-induced global warming who have careers, reputations and funding based entirely on the computer-predicted potential of for catastrophic human-induced global warming. The climate industry is not independent. Evidence for human-induced climate change comes from computer models which cannot repeat what we have observed in nature and they invariably give the result sought. Planet Earth is an ever-changing system and the forces of nature do not obey simplistic computer models.

I am sceptical when postulated temperature increases of 6.4°C (IPCC, 2001) and 4.5°C (IPCC, 2007) from the doubling of atmospheric carbon dioxide ignore the past. If the same methods are used, then at the times in the past when there was a very high carbon dioxide content in the atmosphere and an explosion of life, the atmospheric temperature should have been between 78 and 108°C. This was not the case. Computer models used for predictions of a future fry-and-die scenario have been shown time after time to be wrong.

I have difficulty in believing that the IPCC has anything to do with science when it admits that it used a Greenpeace campaigner to help write an "impartial" report on green energy suggesting that 77% of the world's energy in 2050 could come from sea breezes and sunbeams. The report was not science. It was recycled Greenpeace campaign material. Greenpeace has a pathological hatred of coal-fired power stations and of hydroelectric and nuclear power so the books had to be cooked to show that unreliable uneconomic forms of power will save the world. If one is to trust the IPCC, they need to have no campaigners, advocates or activists as lead authors of IPCC reports. They should not use sources such as newspaper reports, the grey literature or campaign material as references for their reports. No authors should be assessing their own work and press releases and Summaries for Policymakers should be released at the same time as the full scientific reports to enable scientists to evaluate the basis for conclusions. As with any other organisation, there should be a transparent policy on conflict of interest. I am very sceptical when Climategate star Phil Jones cites his own work in the IPCC 2007 report:

> Studies that have looked at hemispheric and global scales conclude that any urban-related trend is an order of magnitude smaller than decadal and longer time-scale trends evident in the series (e.g. Jones et al., 1990; Peterson et al., 1999).

Jones simply ignores the abundant literature that shows an urban heat island effect of up to 11°C in mega cities like New York and that an urban-related trend is not indicative of the global temperature. As the Climategate emails showed, Jones does everything in his power to silence critics or those that try to publish contrary views.

It is little wonder that a recent *Scientific American* poll showed that 81% of those surveyed thought that the IPCC is corrupt with group-think with a political agenda, 77% said that they did not want to pay anything to stop catastrophic climate change, 75% thought that climate change is caused by solar variation or natural processes, 65% thought that we should do nothing about climate change as we are powerless to stop it, 65% thought that science should be kept out of the political process and only 21% thought that climate change was due to human emissions of greenhouse gases. This was an October 2010 survey comprising 5190 respondents. Readers of *Scientific American* could hardly be called scientifically illiterate.

The views of German ex-Chancellor Helmut Schmidt echo the *Scientific American* poll. In a speech in early 2011 at the Max-Plank-Gesellschaft, Schmidt stated:

> The documentation delivered so far by the international group of scientists (Intergovernmental Panel on Climate Change) has been met with scepticism, particularly that some of the involved scientists have proven to be frauds. In any case, targets set by some of the governments have turned out to be less based on science, and based more on politics. In my view it is time for one of our leading scientific elite organisations to take a hard, critical and realistic look at the work of the IPCC, and thus explain in a clear manner the yielded conclusions to the public.

Just because we are alive today does not mean that we are changing the planet's climate. Just because we are alive today does not mean that we understand all natural processes. Nature rules. It always has. Although humans may have a slight effect on the Earth's atmosphere,

41

carbon dioxide in the Earth's atmosphere has never in the past driven global climate and there is an absence of convincing evidence to suggest it does now. Human effects are swamped by the enormous natural changes on Earth.

The *Washington Post* wrote:

> The Arctic Ocean is warming up, icebergs are growing scarcer and in some places the seals are finding the water too hot, according to a report to the Commerce Department yesterday from the Consulate at Bergen, Norway.
>
> Reports from fishermen, seal hunters and explorers all point to a radical change in climate conditions and hitherto unheard-of temperatures in the Arctic zone. Exploration expeditions report that scarcely any ice has been met as far north as 81 degrees 29 minutes. Soundings to a depth of 3,100 meters showed the Gulf Stream still very warm.

The report continues:

> Great masses of ice have been replaced by moraines of earth and stones, the report continued, while at many points well known glaciers have entirely disappeared.
>
> Very few seals and no white fish are found in the eastern Arctic while vast shoals of herring and smelts, which have never before ventured so far north, are being encountered in the old seal fishing grounds.
>
> Within a few years it is predicted that due to the ice melt, the sea will rise and make most coastal cities uninhabitable.

This article was dated 2nd November 1922 during a warm period that may have exceeded any since that time. But the predictions of disaster failed to eventuate. We've heard it all before. It takes a certain strange personality to make predictions of doom and gloom that are continually wrong and we see it again and again today with those in the climate industry. Hundreds of years ago, there were predictions that there would not be enough resources to sustain an increased population. These were wrong. The Club of Rome and many members of the Royal Society in the early 1970s predicted the same. They were wrong. The most unsuccessful doom-and-gloom predictions were by Paul Erlich. In the 1970s, he predicted that China and India would suffer famines and by 1985 hundreds of millions of people would die. They did not and both countries are now self-sufficient in food. He predicted:

> By the year 2000, the United Kingdom would be a group of impoverished islands, inhabited by some 70 million hungry people....If I were a gambler, I would take even money that England will not exist in the year 2000.

He was hopelessly wrong.

On 24th June 1974, *Time* magazine warned us of the forthcoming ice age. We were all going to freeze and die because human-produced aerosols would block sunlight and heat reaching the Earth's surface. They were wrong. The same prominent figures in the human-induced global warming industry were just as prominent in the human-induced global cooling industry of the 1970s. They were wrong. We now get scientific predictions from eminently unqualified souls like Ross Garnaut and Nicholas Stern telling us that the science is beyond reasonable doubt and that we are going to fry-and-die. The only thing beyond reasonable doubt is that these predictions, like all catastrophic predictions in the past, will be wrong.

We fragile humans seem to learn nothing from the repeated failures of the prophets of doom. In 1830, Thomas Macaulay wrote: "On what principle is it that when we see nothing but improvement behind us, we are to expect nothing but deterioration before us."

A short history of planet Earth

Our planet is a wet warm volcanic planet. It formed on a Thursday 4,500 million years ago by the condensation and recycling of star dust associated with the formation of an exceptionally stable star in a good galactic address. Early in the history of the Solar System, Earth was bombarded by massive asteroids. Every time a primitive sea formed by condensation of volcanic steam, it was vapourised by impacting asteroids. This was the limitation to the formation of life on Earth. As soon as the surface and atmosphere of the Earth had cooled and asteroid bombardment decreased, rainwater accumulated and life formed. The evolution of life is inextricably linked to a very weird molecule. It is water. Without water, there would be no volcanoes, no recycling of crustal rocks, no oceans, no climate change and no life.

First life on Earth formed at least 3,800 million years ago and by 3,500 million years ago, this bacterial life had colonised into stromatolites. These bacterial colonies are still with us. Bacteria were the first life on Earth, bacteria are still the dominant life on Earth, bacteria have survived all the natural catastrophes on Earth and they are the largest biomass on Earth. The top 5 kilometres of the Earth's crust contains greenhouse gas-emitting bacteria and we are nowhere close to understanding the role of the greatest biomass on Earth. It was not until Earth was middle-aged that the first great ice ages occurred. For less than 20% of time, Earth has had ice. The first ice ages occurred when the continents were clustered around the equator. The Earth was covered in ice and ice sheets which occurred at sea level at the equator. At that time, ice ages were composed of alternating cold (glaciation) and warm (interglacial) periods when ice sheets grew or melted. It is the same today.

Around 2,500 million years ago a number of irreversible events took place. The continents became thicker, nutrients were washed from retreating ice sheets to fertilise the oceans, bacterial life diversified and the atmosphere started to accumulate oxygen excreted from bacteria. At that time, the atmosphere contained some 30% carbon dioxide and there was a mass extinction of prokaryotic bacterial life killed by oxygen. Although carbon dioxide content was far greater than now and the Sun was slightly fainter, this was not enough to keep the planet warm. Other mechanisms must have operated. Maybe methane was more abundant. Maybe there were far more high-level clouds. Where did all this carbon dioxide go? Large colonies of algae extracted carbon dioxide from the atmosphere and built algal reefs. The process of massive sequestration of carbon dioxide from the atmosphere began. Algal reefs are composed of limestone, a rock that contains 44% carbon dioxide.

Sediments (and sedimentary rocks) contain large amounts of carbon compounds. Limestone is one of these sedimentary rocks. It is still forming. Natural carbon dioxide sequestration is a very rapid process and has been taking place for billions of years. If there were not natural recycling of water and carbon compounds, then the atmosphere would be full of carbon dioxide. It is not. Over this very long period of natural sequestration, atmospheric carbon dioxide was reduced from 30% to less than 0.04%. Carbon dioxide is now a trace gas in the atmosphere and the humans add only trace amounts to that total. The planet has been looking after itself for billions of years without the help of humans.

A giant supercontinent, Rodinia, was fragmented 830 million years ago. Again, the continents were at low latitudes, ice sheets were at sea level at the equator, there were modest sea level rises and falls of 1500 metres. The oceans became fertilised by fine mineral particles washed into oceans by melt waters from retreating ice sheets. Again, there was a diversification of life and, because the oxygen content of the atmosphere had increased, multi-cellular marine animals formed (Ediacaran fauna). The Ediacarans grazed on sea floor algal mats and 542 million years ago, multi-cellular animals developed shells, skeletons and protective coatings because there were enough nutrients in the oceans for muscle functions. These marine animals extracted carbon dioxide from the oceans to make shells and reefs. It took 20 million years for most of the major animal groups to evolve. These armoured animals preyed on the Ediacarans and eventually exterminated them.

These were remarkable times. The atmosphere had a huge amount of carbon dioxide and extensive shallow water platforms of carbonate rocks formed. Some 2,500 times the present atmospheric quantity of carbon dioxide was extracted from the atmosphere and sequestered into rocks. And there it stays. If there were no natural sequestration of carbon dioxide, then this trace gas in today's atmosphere would be a major gas. Massive carbonate platforms where carbon dioxide had been sequestered now became common and exist in all younger geological sequences.

Some 50 million years later, land plants appeared. They have only been on Earth for 10% of time. Land plants accelerated the sequestration of carbon dioxide from the atmosphere. Before then, carbon dioxide was only sequestered into limestone reefs, shells and sediments. The first known major mass extinctions of complex life took place 440-420 million years ago, but life quickly recovered and filled the vacated ecological niches. There was a minor ice age after which life thrived. Another major mass extinction of life occurred 360 million years ago, and again life quickly recovered and plants started to thrive. As a result, the atmospheric carbon dioxide content decreased, the methane content increased and the oxygen content increased to the point where it was common for the Earth's atmosphere to spontaneously ignite. The additional carbon dioxide in the atmosphere was sequestered into plants that later became the Northern Hemisphere coals (Carboniferous) and Southern Hemisphere coals (Permian). It was also sequestered into carbonate platforms that are common in Northern Hemisphere rocks of this age. Again, there was another ice age. Some 251 million years ago there was a mass extinction of 96% of all species when complex life on Earth nearly almost disappeared.

This mass extinction probably resulted from sulphurous gases released from Siberian volcanoes. Sulphurous gases are scrubbed out of the atmosphere by water vapour, acid is formed and acid rain would have killed vegetation and shallow marine life thereby creating a collapse of ecosystems.

Earth recovered back to being a normal wet warm planet, only to be interrupted by another mass extinction of life some 217 million years ago, possibly from asteroid impaction. A giant supercontinent started to fragment and the Atlantic Ocean started to form. Extension of the Earth's continental crust to form an ocean resulted in a huge degassing into the atmosphere of water vapour, carbon dioxide and methane associated with monstrous outpourings of submarine volcanic rocks. Plant life thrived with the extra atmospheric carbon dioxide and carbon dioxide was again locked away by the normal suspects (limestone, coal, carbon-rich sediments). Earth again recovered from this mass extinction, life continued to diversify and the continents continued to drift. Sudden warming events took place 183 million and 120 million years ago. The first may have been due to a massive injection of basalt into the Karroo Basin of South Africa. These melts would have released large quantities of carbon dioxide and boiled off methane from marine sediments. This methane would have quickly oxidised to carbon dioxide. Australia 120 million years ago was enjoying the then warm climate of the South Pole, which was exploited by dinosaurs. There were occasional volcanic eruptions and every now and then alpine glaciers surged to the sea and calved icebergs. About 100 million years ago, a great southern continent started to fragment and Australia started to move northwards at a few centimetres a year. It still is.

The mass extinction 65 million years ago is the one we all know about. The story is that a giant asteroid hit Mexico, vaporised the target site and sulphur gases filled the atmosphere. These gases were scrubbed out of the air by water vapour, acid rain fell, plants and floating marine animals died and there was a total collapse of the planet's ecosystems. What a wonderfully exciting evocative story: a huge asteroid with "Death to all Dinosaurs" written on it colliding with Earth. The problem is that the science on this matter is not settled. New information suggests that some 300,000 years later there were massive sulphur dioxide-emitting volcanoes in India that created acid rain... you know the rest of the story. The dinosaurs may have become extinct from a totally different event 300,000 years after an asteroid hit Mexico. There is a raging argument in the scientific literature about this mass extinction. This is why, as a scientist, I do not accept statements such as "the science is settled" because science is never settled. Statements on climatic matters by politicians and advocates that "the science is settled" have nothing to do with science and advertise another agenda. Such statements are an attempt to dictate how we think and avoid healthy debate.

Some 55 million years ago, an extraordinary event occurred on Earth. A major emission of methane occurred. This may have derived from the injection of a large amount of basalt off the Norwegian coast. The basalt could have boiled off carbon dioxide and heated marine sediments would have released methane that would have quickly oxidised to carbon dioxide. At present, there are huge volumes of molten basalt beneath the Afar Rift in East Africa just waiting to release carbon dioxide into the atmosphere. Alternatively, a cluster of intrusions of the diamond-bearing rock kimberlite may have intruded from deep in the Earth at that time. Kimberlite ascent is driven by rising expanding and cooling carbon dioxide gas that is released into the atmosphere. We don't know why this event took place, there are many competing and contradictory ideas floating around in the peer-reviewed literature and a healthy debate rages. This is the key to peer review. If a paper is peer-reviewed, it does not mean it is correct and it does not mean that it is the one and only accepted explanation of a phenomenon. Peer-reviewed papers date very quickly because new evidence requires new explanations.

Whatever the origin, there was a sudden injection of methane into the atmosphere and the Earth warmed for about 170,000 years. The methane oxidised to carbon dioxide. This warming was not irreversible, was not catastrophic and we humans could have lived through it as average global temperatures probably rose about 6°C and by 10 to 20°C at the poles. The temperature change was less than the difference between summer and winter in temperate climates. Planet Earth recovered without the help of humans. Various models have been used to try to understand how quickly the carbon dioxide could have been removed from the atmosphere. The models show that carbon dioxide removal was far slower than the actual carbon dioxide sequestration.

India collided with Asia 50 million years ago. Local climate was changed and carbon dioxide continued to be drawn down into soils and sediments from the atmosphere as the mountains rose. The Earth's climate has been cooling for the last 50 million years. South America had the good sense to pull away from Antarctica 37 million years ago, a circum-polar current isolated Antarctica from warm water and local ice caps joined to form a single continental ice sheet on Antarctica some 34 million years ago. We are in another ice age with alternating glaciations and interglacials. At present, we are approaching the end of an interglacial and, unless we can change the behaviour of the Sun and the Earth's orbit, we will enter the next inevitable glaciation. The Antarctic ice sheet has waxed and waned for the last 34 million years. As planet Earth cooled, the slight cyclical variations in the Earth's orbit and distance from the Sun started to have a profound influence on climate. However, there were some short sharp periods of global warming unrelated to carbon dioxide or industry (as humans were not around then). Climate changes drove human evolution over the last 5 million years. In south-eastern USA, between 5.2 and 2.6 million years ago, atmospheric carbon dioxide content was more than now as were global temperatures (2 to 3°C) and sea level (10 to 25 metres). This was probably driven by the regular changes in the Earth's orbit. With the closure at Panama of the connection between the Atlantic and Pacific Oceans 2.67 million years ago by volcanoes, Earth started to cool. Coincidentally, there was a supernoval eruption at the same time. The bombardment of the Earth by cosmic rays from this supernova eruption led to the formation of low-level clouds that cooled the Earth's surface. The two coincidental processes accelerated cooling. As a result, the Greenland ice sheet formed.

Climate fluctuated in cycles between warm and cold periods every 41,000 years. This was driven by changes in the Earth's axis. About 1 million years ago, the climate started to fluctuate between cold and warm periods on 100,000-year cycles driven by changes in the orbit from elliptical to circular patterns. We are currently in the warmer interglacial phase of an ice age that has been in progress for 34 million years. During the last interglacial, sea level was 4 to 9 metres higher than now and temperatures were 3-5°C higher than now. During the current interglacial 6,000 years ago, sea level was 2 metres and temperature 2°C higher than now. We do not know when the current ice age will end. However, we cannot escape the fact that the current interglacial will end and we will enjoy another 90,000 years of glaciation. Previous glaciations had kilometre-thick ice sheets that covered Canada, northern USA, most of the UK, most of Europe north of the Alps, most of Russia, all of Scandinavia and elevated areas in both hemispheres. There is no reason why the next inevitable glaciation will be any different.

This chronicle of our planet is an evocative story underpinned by empirical evidence. No computer models have been able to replicate this chronicle. If computer models cannot look backwards, then they cannot look forwards to predict future climate changes. With new evidence, this chronicle is refined so the story of our planet is never settled science. A huge number of very large forces have interacted to produce this chronicle. Some forces were random, others were cyclical and others were irreversible. Much is still unknown. Many past climate changes have been greater and more rapid than any changes measured today. If we humans wanted to change climate on Earth, we would have to stop bacteria doing what bacteria do, change ocean currents, manage the drift of continents, change the Earth's orbit, control the variability of the Sun and control supernoval eruptions. Piece of cake.

Planet Earth is dynamic. It always has been. Climate has always changed and to call someone a climate-change denier advertises the fact that the name caller knows nothing about the history of the Earth. Of course climate changes, no one denies that. Past climate cycles have been of galactic (143 million years), orbital (100,000, ~41,000 and ~21,000 years), solar (1,500, 210, 87, 22 and 11 years), oceanic decadal (~30 years) and lunar tidal (~18.6 years) origin. Oceans have decadal oscillations (e.g. Pacific Decadal Oscillation) and apparently non-cyclical major events (e.g. El Niño-La Niña) where the surface temperature of ocean water changes. Some of these phenomena have only just been discovered and there is still much debate about the role of electromagnetic fields, gravity, the Sun, cosmic radiation, volcanoes and ocean oscillations. All are important because they affect the transfer and balance of heat on the surface of the Earth. The heat capacity of water determines that it is not the temperature of the atmosphere that heats the ocean surface waters. It is the temperature of the ocean surface that heats the atmosphere. The oceans contain more than 50 times as much heat as the atmosphere. A good analogy is in the bathroom. A bath full of hot water heats the air in the room. Conversely, heated air in the bathroom does not heat a bath full of water. So too with the atmosphere and oceans. The 3,000 ARGO buoys in the world's oceans are showing that the surface temperature of the ocean is decreasing yet the carbon dioxide content of the atmosphere is increasing. This is commensurate with the 21st century cooling (derived from satellite and land measurements) during a period of increasing carbon dioxide. This shows that there is no relationship between increasing atmospheric carbon dioxide from human emissions and global warming. There must be other powerful driving forces.

Galactic climate cycles derive from increased bombardment of the Solar System with cosmic rays. These cosmic rays induce the formation of low-level clouds that reflect heat. The Earth then cools. The six major ice ages that planet Earth has enjoyed were at times when Earth was in the Sagittarius-Carina (twice), Perseus, Norma, Scutum and Orion arms of the galaxy. Wobbles in the Earth's orbit produce cycles of warm and cold (Milankovich cycles) resulting from changes in the distance between the Earth and the Sun. The Sun does not have a constant emission of energy. It has a number of regular cycles and random outbursts of energy. These influence climate because they result in changes in the solar magnetic field that, in turn, protects the Earth from cosmic ray bombardment. It may appear heretical to those advocating human-induced climate change derived from traces of carbon dioxide, but that great ball of energy in the sky we call the Sun actually drives surface energy, climate systems and life on Earth.

In the past, climate changes have sporadically occurred as a result of super volcanoes, supernoval eruptions, moving continents, opening and closing of seaways and possibly impacts. Large volcanic eruptions at tropical latitudes (like Tambora, Indonesia 1815; Krakatoa, Indonesia 1883) eject aerosols into the stratosphere. These aerosols reflect light and heat and produce cooler stormy weather that lasts for a few years. They also eject sulphurous gases. Terrestrial super volcanoes (such as Yellowstone, USA; Taupo, NZ: Kamchatka, Russia) eject thousands of cubic kilometres of aerosols into the atmosphere and can have a profound cooling effect. Planet Earth started to cool 116,000 years ago at the beginning of the most recent glaciation. During the cooling, Toba (Indonesia) erupted 74,000 years ago. It filled the atmosphere with aerosols and accelerated the rate of cooling. Humans very nearly became extinct. Aerosols also fill the atmosphere after an asteroid impact.

The collision of India with Asia 50 million years ago also resulted in the uplift of mountains. These new mountains were stripped of soils, new soils formed, river sediments were dumped in the adjacent plains and bare rock in alpine areas created monsoonal updrafts. The formation of soils results in the extraction of carbon dioxide from the atmosphere.

The climate system is like an orchestra. The conductor of the orchestra is the Sun. Water vapour is the first violins and carbon dioxide the second violins. The second violins follow the first violins and the whole orchestra follows the conductor. Yet we are being told that the second violins instruct the conductor and the music we hear from the orchestra only comes from the second violins. Climate, like orchestral music, is very complex. A blast from the brass may represent sudden events such as an asteroid impact or supernoval eruption and the woodwinds come in and out of influencing the music and these are represented by volcanoes and continental and ocean changes. Layering in orchestral music is like the cycles of climate. Orchestral music often tells us what will be coming. So too with climate cycles.

What warming?

The facts have not changed. Computer models show that the tell tale signature of human-induced global warming would be a hot spot high in the atmosphere above the equator. Weather balloons have scanned the skies for decades and no hot spot exists. Ice core data showed that temperature rises were mirrored by rises in the atmospheric carbon dioxide content on a coarse scale. More detailed finer scale analyses showed that atmospheric temperature rises at least 800 years before the rise in atmospheric carbon dioxide. Warmists claim that even if carbon dioxide does not start the warming trend, it amplifies it. However, the past shows that for most of time the atmospheric carbon dioxide content has been far higher than now and we have not had runaway global warming. This means that such amplifications have not occurred in the past and just happen to occur now.

Another inconvenient question is: If carbon dioxide amplifies the warming trend, why is it a fact, as demonstrated time and time again by the ice core records, that the warming trends reached the peak of the cycle then start to decline while the carbon dioxide content continued to increase? If carbon dioxide drives global warming, then this is not logical.

Warming caused by carbon dioxide was predicted from computer models and how could they possibly be wrong? Something other than carbon dioxide caused the warming recorded in ice cores. Satellites circling the planet all the time have shown that there has been no increase in warming since 2001. However, atmospheric carbon dioxide continues to increase so clearly temperature is unrelated to atmospheric carbon dioxide content. Carbon dioxide has done its job. Any further increases in atmospheric carbon dioxide will produce an increasingly smaller temperature increase.

We have been told time and time again for the last 20 years that we all will fry-and-die. And that it was our fault. We were going to have longer and more intense droughts. There were going to be more big forest fires, more floods, more frequent and more intense cyclones and tornadoes. The Great Barrier Reef was going to be bleached and die. The ice sheets were going to collapse, the area of sea ice was going to decrease and winter snow in many parts of the Northern Hemisphere was to be a very unusual event. We were told that sea level would rise rapidly and that we had better head for the hills. The oceans were going to become so acid that all those invisible lovable critters would die. There were going to be tens of millions of climate refugees, island states were to be inundated, pandemics would wipe out the millions who had not already starved from food shortages and the predicted problems were so catastrophic that democratic processes would have to be suspended. Immediately. Computer predictions told us of this scary future.

We are told that an overwhelming number of scientists supported this vision for the future although we are occasionally told that there are a few nutters with vested interests, the climate change deniers, who do not follow the party line. But we could forget them, as they are supported by big oil. In fact, the only contact I have with big oil is when I fill up the tank in my car. I pay big oil, they don't pay me. Maybe I should ask for a discount for services rendered as fuel is getting pretty expensive now.

Big government outspent big oil. The US government spent $30 billion on looking for a climate crisis. And what did this $30 billion find? Nothing. Globally, although $100 billion has been spent looking for a human signal of climate change. It has not been found hence any warming is so slight that it cannot be detected. With so much money sloshing around, no wonder crooks, banks, traders and bottom-feeders swim around advocating emissions trading and carbon taxes. Your money is easy pickings. If it cannot be shown that human emissions drive climate change, then there should be no further funding. The climate is OK.

Nature has a great sense of humour and so far every single one of these scientific predictions has turned out to be wrong. Some of them are so hopelessly wrong that it is hilarious. It is no mean achievement to be wrong on everything, you'd think the climate industry could have jagged the correct answer for one single prediction. They didn't. We have not fried and died. We have warmed and cooled. More people still die from extremes of cold than extremes of heat. There have been greater droughts, bushfires, floods and cyclones in the past than in recent times. The Northern Hemisphere has had a couple of bitter winters with metres of snow which have collapsed transport systems and a couple of cool summers. Some areas have had a very slight increase in sea level, others have had a decrease. There has been no widespread inundation by the sea. Island states such as the Maldives and Tuvalu still exist and are actually increasing in size. Above sea level. The oceans have not become acid, coral reefs have not dissolved or disappeared and life keeps doing what life does. Despite a great human population increase over the last 20 years, we have not seen tens of millions of climate refugees, pandemics and mass starvation. What we have seen is decreasing disease, less poverty and more food.

We are told that human-induced global warming is all our fault yet there is a glimmer of hope. If we in the West pay huge carbon taxes, trade the rights to emit carbon dioxide, completely change our economy, reduce our standard of living to that of the Third World and give away our freedoms, then we will survive. This sacrifice is to be made by us, not by those telling us how to live our lives. We are not told the bad news: the costs for such action will bankrupt us.

Even the most draconian global carbon reduction proposals by Australia would only lower global temperature in 2050 by 0.01°C. This is huge pain for no gain. If I stand up, or go from one room to another or walk outside, the temperature change is far more than 0.01°C. The daily and seasonal temperature variation is hundreds of times greater than this. Any action to lower global temperature by 0.01°C is totally pointless.

Many in the Northern Hemisphere take holidays in warmer climates and retired people commonly move to a warmer climate. These are the real climate refugees. The Europeans wanted to take drastic action to ensure that the computer-predicted global temperature rise was kept to 2°C. Imagine if the Northern Hemisphere warmed by 5°C, as it did in the Roman and Medieval Warmings. Europe would have an agricultural boom. Northern Canada, Greenland and Siberia would again be habitable, crops could again be grown at high latitudes and economies would again thrive.

Over the last 25 years, there has been no observational evidence produced to show that human emissions of carbon dioxide create global warming. This is despite intense investigation that cost tens of billions. After such an effort and cost, if there was something to be found, then it should have been found. However, only contrary evidence has been discovered. In 2006, Henrik Svensmark experimentally validated astronomical and isotope measurements that suggested solar energy reaching the Earth is very variable and is related to changes in the Sun's activity, extraterrestrial bombardment by cosmic rays and the resulting changes in cloudiness. Experiments by Svensmark showed that cosmic rays influence cloud formation, that low-level clouds have a cooling effect on Earth and that over the last few decades, there have been fewer clouds than normal because of the Sun's magnetic field. Independent experiments at CERN validated Svensmark's work. Solar physicists have shown that the 20th century was an active time for the Sun and that the Earth's climate is far more sensitive to very slight changes in the Sun than assumed by the IPCC. The Sun is now quietening and some physicists are suggesting that we could be facing cold times. There are now three levels of experiments that seem to support the solar-cosmic ray hypothesis for cloud formation yet the IPCC ignores this work. The IPCC would have no reason to exist if it were accepted that clouds and not carbon dioxide were a major driver of climate.

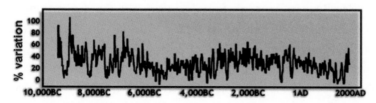

Figure 3: *Changes in isotopes (C^{14}, Be^{10}, Al^{6}, Cl^{6}. Ca^{41}, Ti^{44}, I^{129}) formed by cosmic rays over time with variations resulting from changes in the cosmic ray-shielding solar magnetic field or variations in the amount of cosmic rays striking Earth.*

Climate is very complex. Science continues to make new discoveries which show this complexity and the secondary role of carbon dioxide. However, for those in the climate industry to re-evaluate their beliefs will be hard. There is too much cash at stake.

Follow the money

There are a huge number of jobs at stake if the theory of global warming from human emissions of carbon dioxide is wrong. Think of all those bureaucratic jobs, research grants and scientific careers that would just evaporate. Think of all the banks, trading houses, investment groups and energy companies that would miss out on skinning us alive.

The most prominent and radical green activist groups are partially funded by taxpayers.[4] The same groups try to stop the economic development that employs people and provides funds for government by creating a taxation base. While State governments are trying to build dams and coal-fired power stations to keep the population fed and watered, green groups are using moneys given to them by State governments to object to the building of infrastructure. Gilding the lily is the strong suit of green groups. For example, Friends of the Earth claimed on its web site that it receives no government or corporate funding. But in fact they received $55,000 from the Victorian Department of Sustainability and the Environment as well as annual recurrent funds. Many other groups have had their snouts in the taxpayer trough since 2006 for a total of $10.78 million: Australian Conservation Foundation ($2.9 million); Conservation Council of SA ($0.85 million); Conservation Council of WA ($0.28 million); Environment Defenders Office ($1.2 million); Environment Victoria ($4.04 million); Friends of the Earth ($65,000); Queensland Conservation Council ($0.87 million); Total Environment Centre

---

4 Chris Kenny, *The Australian*, March 19th, 2011. *Taken for granted: how tax dollars are helping to fund green agendas.*

($0.45 million) and the Wilderness Society ($0.125 million). Ask your local politician why taxpayers' money is being used to help unelected pressure groups put people out of work. Put pressure on your local politician.

The Federal Department of Climate Change and Energy Efficiency has sprayed around huge amounts of money. Senate papers show the grants administered. Climate boffins at the University of Tasmania received $306,697 plus $17,444 for Nathan Bindoff to attend the IPCC Fifth Assessment Report Scoping Meeting, $110,000 for Australian participation in the IPCC Fifth Assessment Report Scoping Meeting and, together with the Australian Academy of Sciences, $39,050 to find out the views of planners, local governments and engineers. Another $6,667 went to Ian Allison to attend a workshop on sea level.

Canberra institutions scored exceptionally well from the Canberra-based Department of Climate Change and Energy Efficiency. The Australian Carbon Trust Limited only received $6 million and a top up of a cool $94.908 million of taxpayers' money. The Australian Bureau of Agricultural and Resource Economics received $55,000 for a chosen one to attend the IPCC Fifth Assessment Report Scoping Meeting and the CSIRO received $3,152,422 comprising many grants with a few bedfellows (Engineers Australia, various local councils) as well as grants of $7.748 million, $361,876, $1,018,225, $990,000, $25,000, $539,000, $258,544, $208,000, $35,200, $660,000, $770,000, $5,536,300, $850,000, $405,000, $550,000, $200,000 and $319,000. A grant of $17,443 was give to John Church to attend the IPCC Fifth Assessment Report Scoping Meeting. These funds were above and beyond what the CSIRO is given from other areas of the Australian government.

The Australian Academy of Science received $227,150 for climate studies. When the Australian Academy of Science makes a public statement on global warming, it should be taken with a very large pinch of salt. None of their statements declare that they have a financial interest in the matter. Geoscience Australia received $561,000 to look at wind and $795,000 for groundwater research in East Timor. The Australian National University received grants of $523,217, $249,609 and $94,050 as well as two separate grants of $121,000 and $8,666 for those in the climate industry to attend the IPCC Fifth Assessment Report Scoping Meeting. Kurt Lambeck from the ANU received $55,000 as support for Australian participation in the IPCC Fifth Assessment Report. The Australian Defence Forces Academy also received $55,000 for attendance at the IPCC Fifth Assessment Report Scoping Study. Besides getting funding from other areas of government, the Australian Bureau of Meteorology received $2.657 million to play modelling games with the CSIRO as well as $2.265 and $1.780 million. A grant of $132,000 to the CSIRO was for the chosen few to attend the IPCC Fifth Assessment Report Scoping Meeting.

In Sydney, Macquarie University scored $62,370 for participation in the IPCC Fifth Assessment Report Scoping Meeting. The University of New South Wales received $165,000 to set up a web-based climate system, $145,000 for author support for the IPCC Fifth Assessment Report and another two tranches of $17,443 for a warmist to attend the IPCC Fifth Assessment Report Scoping Meeting. In addition, the web site of the Climate Change Research Centre at the University of New South Wales showed some $5.238 million of grants over the last few years from other government sources. This is only a fraction of the funds required to keep this centre operating but it gives the picture.

Monash University received $305,419 and, on top of this, only $55,000, $55,000, $17,443 and $17,443 for various warmists to attend the IPCC Fifth Assessment Report Scoping Meeting. RMIT received $55,000 for someone to attend the IPCC Fifth Assessment Report Scoping Meeting, $88,569 for simulation and another $578,000 to look at sea ports. The University of Melbourne scored $123,035 for rainfall estimate work and also was given $55,000 and $17,443 for climate alarmists to attend the IPCC Fifth Assessment Report Scoping Meeting. David Karoly received funds for the IPCC Fifth

Assessment Report Scoping Meeting, and Senate parliamentary records of June 2009 reveal he also received $1.9 million in grants from the federal government for climate change research. The total of grants awarded by the government to the University of Melbourne is many millions of dollars. An exact figure is difficult to determine because such grants derive from various government departments. Those listed in the Senate papers of June 2009 showed $7,140, 360 went to the University of Melbourne. The figure now would be well over $8 million. What did the taxpayer get for this money? Victoria University received $220,000 for one project and $55,000 for one of their folk to be supported for the IPCC Fifth Assessment Report.

The University of Queensland received grants of $72,000 and $17,443 for Ove Hoegh-Guldberg to attend various IPCC meetings. James Cook University received $440,000, $311,714, $122,845 and $166,000 with a grant of $6,000 for a warmist to attend the IPCC meeting. Griffith University received only $133,700 for various warmists to attend IPCC Fifth Assessment Report Scoping Meeting. Murdoch University received $55,000 for one of their warmists to attend the IPCC Fifth Assessment Report Scoping Meeting; so did Curtin University. The WA Department of Environment and Conservation scored $110,000.

The Australian Climate Change and Business Centre Limited received $52,500 to assist with a couple of conferences. The conference programs show that only those in the inner sanctum of the climate industry presented papers. Another NGO, the Climate Institute, received only $70,000 for modelling the carbon price. A representative from a Norwegian alternative energy group received $60,500 of Australian taxpayers' money to attend the IPCC Fifth Assessment Report Scoping Meeting. A senior executive of Munich RE received $63,370 as support to participate in the Fifth IPCC Assessment Report. One might ask why the Australian government is giving grants to private corporations that make a profit from the global warming scare campaign. A power company with a tiny 1-megawatt solar power station at Alice Springs was gifted $3,639,836. Why should the Federal Department of Climate Change and Energy Efficiency give money to inefficient power companies? Surely they can stand on their own feet and, if they are operating inefficiently and require a subsidy, they should be allowed to go broke. Other inefficient businesses go broke so why shouldn't a power company?

Various other interesting names appear. Why did the Federal Department of Climate Change and Energy Efficiency give a grant of $22,000 to the lawyers Baker McKenzie to write a set of rules? Why do various associations and NGOs get a slice of the cake (Engineers Australia, Australian Green Council Infrastructure, Australian Council for Social Services, Australian Institute for Landscape Architects, Australian Red Cross, Climate Change and Business Centre Ltd)?

Why do various local governments get money for climate change (Alice Springs Town Council, City of Geraldton-Greenough, Shire of Chapman Valley, Shire of Northampton, Shire of Irwin, Gloucester Shire Council, Upper Hunter Shire Council, Muswellbrook Shire Council, Singleton Shire Council, Dungog Shire Council, Cessnock Shire Council, Greater Taree Shire Council, Maitland City Council, Belyuen Shire Council, Walgett Shire Council, Coomalie Shire Council, East Arnhem Shire Council, West Arnhem Shire Council, Tiwi Islands Shire Council, Nambucca Shire Council, Bellingen Shire Council, Kempsey Shire Council, Towong Shire Council, Alpine Shire Council, City of Wodonga)?

These are the grants administered by the Federal Department of Climate Change and Energy Efficiency as of 1st January 2009. It adds up to over $142 million and I have only been able to find a fraction of the money thrown at the climate industry. What is presented here is a gross underestimation but it gives the picture. We do know that the Department of Climate Change and Energy Efficiency has been even more generous and recently gave $180,000 per year to Tim Flannery for three days a week of public appearances for a few years telling us all how a "Carbon Tax" is good for us. And Ross Garnaut and his staff certainly don't come cheaply.

What has been shown time and time again is that those who get huge amounts of public money from the Department of Climate Change and Energy Efficiency do not want to appear in a public debate with scientists who might actually know something about climate. There is only one exception. Ben McNeil (UNSW) participated in the *Spectator Australia* debate in August 2011 but actually did not talk about science. Frightening taxpayers is a big business from which the climate industry benefits hugely. And who in the Department of Climate Change and Energy Efficiency makes grant decisions? This department is not filled with scientists and engineers at the peak of their research careers. It is filled with bureaucrats who probably have a very scant knowledge of the science of climate. God knows how much money they fritter away now or what moneys come from other Federal departments and state departments go to the climate industry. The figure of $142 million does not include the $100 million wage bill to keep the Department of Climate Change and Energy Efficiency afloat. There is almost no grant money for acquisition of new data. Money is for jaunts, creation of models, building of web sites and gifts to industry. So at least $142 million is used to massage someone else's data.

When you read of criticisms of *Heaven and Earth* and this book, there is one simple number to keep in mind. $142 million. The climate industry, academies, professional societies, NGOs, insurance companies and alternative energy companies are not engaged in independent transparent research to add to knowledge or to satisfy intellectual curiosity. Every single one of my few noisy scientific critics is either a direct recipient of huge amounts of money or is in an institution that receives masses of cash because they support the government ideology. The same groups that claim that those scientists who object to their lack of science underpinning climate activism are funded by "big oil". Not one institution that receives grant money dares to question the popular idea that human emissions drive global warming. Clearly no one in the Department of Climate Change and Energy Efficiency asks: What if we are wrong? The basis of science is to question. This is not done by those in the climate industry.

No wonder I am attacked, I am a party pooper and could reduce the cash flow of the climate industry. We taxpayers are having huge amounts of our own money spent to frighten us. What is even more interesting is that not one single climate contrarian received a penny from the Department of Climate Change and Energy Efficiency or the Australian Research Council. Not one. The 1,136 Australian Research Council grants total around $392 million. Some 10% of these mention climate change, a sure fire winner as this is the way to get funded.

Freedom of information requests in the US show that James Hansen, the guru of human-induced global warming, does not deal with small change. Hansen once claimed that *"coal-fired power stations are factories of death."* In a lawsuit, it was claimed that Hansen shared a $1 million prize from the Dan David Foundation for his "profound contribution to humanity." He would have received anything from $333,000 to $500,000. He received the 2010 Blue Planet prize worth $550,000 from the Asahi Glass Foundation, which recognises efforts to solve environmental issues, the $100,000 Sophie Prize for his "political activism," worth $100,000, speaking fees totaling $48,164 from a range of mostly environmental organizations, a $15,000 participation fee, waived by the W.J. Clinton Foundation for its 2009 Waterkeeper Conference and $720,000 in legal advice and media consulting services provided by the George Soros Open Society Institute. In 2010, his in-kind travel expenses were $59,750 with $26,000, $18,000 and $7,000 for travel to Australia, Japan and Norway, respectively for Hansen and his wife. In a period of five years, Hansen earned in outside income between $1.47 million and $2.67 million, in addition to his basic salary as a government employee of $180,000. Under the terms of contract governing that salary, Hansen is forbidden from privately benefiting from public office and from taking money for activities related to his taxpayer funded employment.

Not one great scientific discovery has ever been made by consensus. They have all been made by researchers going out on a limb and challenging the popular contemporary knowledge. The grant-awarding political policies of the Department of Climate Change and Energy Efficiency and the Australian Research Council guarantee that there will be no great discoveries in Australia about climate. The Department of Climate Change and Energy Efficiency appears only interested in those that support the current political dogma and they are not genuinely interested in either the environment or energy efficiency. This does not appear to be a wise way to create policy or prepare for the future.

The rot has already set in. A review of the Australian government's failed solar rebate scheme showed that $1 billion of taxpayers' money was wasted on subsidies for household solar panels. This favoured the rich and did little to reduce Australia's greenhouse gas emissions, especially as most equipment was manufactured offshore resulting in sending hundreds of millions of dollars outside the country. Some $295 million worth of Chinese solar panels were imported into Australia.

Corruption, fraud and porky pies

In my opinion, Climategate in late 2009 was a window into the depth of fraud and corruption in the climate industry. The emails show conspiracy, collusion to exaggerate warming data, destruction of information that could be embarrassing, organised resistance to releasing information, data manipulation and private admissions of errors in their public claims. One email even gloated over the death of a contrarian. If this were in the corporate world, bigger gaols would have to be built. Someone hacked into or internally leaked files from the computers of the University of East Anglia's Climate Research Unit and released to the internet 61 megabytes of data comprising

1079 emails and 72 documents. These emails showed the depth of the fraud. They also showed very poor record keeping and huge problems with the computer codes created to analyse the data.

There was an admission of altering of data to hide the fact that cooling is taking place:

> I've just completed Mike's Nature trick of adding up the real temps to each series for the last 20 years (i.e. from 1981 onwards) and from 1961 for Keith's to hide the decline.

The emails showed private doubts, contrary to public scare tactics, about whether global warming is actually taking place and also claim that, when the data is contrary to ideology, then the data must be wrong:

> The fact is that we can't account for the lack of warming at the moment and it is a travesty that we can't. The CERES data published in the August BAMS 09 supplement on 2008 shows there should be even more warming, but the data are surely wrong. Our observing system is inadequate.

There is caution by some of the climate industry that the cake might be over iced:

> I know there is pressure to present a nice tidy story as regards 'apparent unprecedented warming in a thousand years or more in the proxy data' but in reality the situation is not quite as simple. We don't have a lot of proxies that come right up to date and those that do (at least a significant number of tree proxies) some unexpected changes in response that do not match the recent warming. I do not think that this is wise that this issue be ignored in the chapter.

There is no doubt that the climate industry is a cosy little club with leading members Michael Mann and Phil Jones organising gongs for each other. In such a cosy self-adulating club, no one will break ranks.

> *Mann to Jones* (4.12.2007): By the way, still looking into nominating you for an AGU award, I've been told that the Ewing medal

wouldn't be the right one. Let me know if you have any particular options you'd like me to investigate...

Jones to Mann (4.12.2007): As for the AGU – just getting one of their Fellowships would be fine.

Mann to Jones (4.12.2007): Will look into the AGU fellowship situation ASAP.

Mann to Jones (2.6.2008): Hi Phil, This is coming along nicely. I've got 5 very strong supporting letter writers lined up to support your AGU Fellowship nomination (confidentially: Ben Santer, Tom Karl, Jean Jouzel, and Lonnie Thompson have all agreed, waiting to hear back from one more individual, maximum is six letters including mine as nominator).

Surprise, surprise, Jones was elected to Fellowship of the American Geophysical Union (AGU) in January 2009. Of course, there is the payback.

Mann to Jones (16.5.2009): On a completely unrelated note. I was wondering if you, perhaps in tandem w/some of the other usual suspects, might be interested in returning the favour this year; I've looked over the current list of AGU fellows, and it seems to me that there are quite a few who have gotten in (e.g. Kurt Cuffey, Amy Clement, and many others) who aren't as far along as me in their careers, so I think I ought to be a strong candidate. Anyway, I don't want to pressure you in any way, but if you think you'd be willing to help to organize, I would naturally be much obliged. Perhaps you could convince Ray or Malcolm to take the lead? The deadline looks as if it is again July 1 this year.

So now you know it. This is the way mediocre people become "eminent". If this is the way the climate industry writes to each other, God knows what they say to each other in private at climate conferences.

Climategate shows the destruction of evidence underpinning the idea of human-induced global warming and removing material that may be subject to a Freedom of Information request:

> Can you delete any emails you may have had with Keith re AR4.
> Keith will do likewise. He's not in at the moment – minor family
> crisis. Can you email Gene and get him to do the same? I don't have
> his new email address. We will be getting Caspar to do likewise.

In the commercial world, such evidence would be seen for what it is. The Medieval Warming has always been a problem for the climate industry and, rather than argue the science, it is far easier to just remove history from the record. Another case of massaging data:

> ...Phil and I have recently submitted a paper using about a
> dozen NH records and fit this category, and many of which are
> available nearly 2K back-I think that trying to adopt a timeframe
> of 2K, rather than the usual 1K, addresses a good earlier point
> that Peck made w/regard to the memo, that it would be nice to
> try to "contain" the putative "MWP", even if we don't yet have a
> hemispheric mean reconstruction available that far back...

As for manipulating data and plots, it is hard to beat the brazenness of Michael Mann. He is fully aware that temperatures have declined in the late 20th century and admits that he is trying to cook the books to show that such cooling does not take place:

> I am perfectly amenable to keeping Keith's series in the plot, and
> can ask Ian Macadam (Chris?) to add it to the plot he has been
> preparing (nobody liked my own color/plotting conventions so
> I've given up doing this myself). The key thing is making sure the
> series are vertically aligned in a reasonable way. I had been using
> the entire 20th century, but in the case of Keith's, we need to align
> the first half of the 20th century w/ the corresponding mean values
> of the other series, due to the late 20th century decline.

There were clear attempts to control the peer-reviewed literature, bully editors, have a mass walk-out of an editorial board, black ban scientific journals that publish science contrary to their views and pressure publishers in the tried and proven ways of union thugs. Anyone with alternative conclusions based on evidence has great difficulty in having their science published by the climate industry mafia. Those who make claims regarding the veracity of human-induced global warming by advocates and politicians should take a deep breath and a cold shower.

> There was the danger of always criticising the skeptics for not publishing in the "peer-reviewed literature". Obviously, they found a solution to that – take over a journal! So what do we do about this? I think we have to stop considering "Climate Research" as a legitimate peer-reviewed journal. Perhaps we should encourage our colleagues in the climate research community to no longer submit to, or cite papers, in this journal. We would also need to consider what we tell or request of our more reasonable colleagues who currently sit on the editorial board...What do others think?

And:

> I will be emailing the journal to tell them I'm having nothing more to do with it until they rid themselves of this troublesome editor.

And:

> It results from this journal having a number of editors. The responsible one for this is a well-known skeptic in NZ. He has let a few papers through by Michaels and Grey in the past. I've had words with Hans van Storch about this, but got nowhere. Another thing to discuss in Nice!

And:

> PS Re CR, I do not know the best way to handle the specifics of the editing. Hans van Storch is partly to blame – he encourages the publication of crap science 'in order to stimulate debate'. One approach is to go direct to publishers and point out the fact that their journal is perceived as being a medium for disseminating misinformation under the guise of refereed work. I use the

word 'perceived' here. Since whether it is true or not is not what the publishers care about – it is how the journal is seen by the community that counts.

And:

> Much like a server which has been compromised as a launching point for computer viruses, I fear that "Climate Research" has become a hopelessly compromised vehicle in the skeptics' (can we find a better word?) disinformation campaign, and some of the discussion that I've seen (e.g. a potential threat of mass resignation among the legitimate members of the CR editorial board) seems, in my opinion, to have some potential merit.

And (on the journal *Energy and Environment*):

> I don't read E&E, gives me indigestion—I don't even consider it peer-reviewed science, and in my view we should treat it that way. i.e., don't cite, and if journalists ask us about a paper, simply explain it's not peer-reviewed science, and Sonja B-C, the editor, has even admitted to an anti-Kyoto agenda!

Scientists who do not normally publish in the climate literature, who are from the field of astronomical observations and who disagree with the climate industry also come in for great respect:

> Might be interesting to see how frequently Soon and Baliunas, individually are cited (as astronomer). Are they any good in their own fields? Perhaps we could start referring to them as astrologers (excusable as…'oops, just a typo').

The attempted abuse of the peer review process shown by the Climategate emails was an attempt to lean on journals, editors and reporters to make sure that critics did not get an airing. This is elegantly put in an email by Phil Jones of the Climate Research Unit of the University of East Anglia:

> We will keep them out somehow – even if we have to redefine what peer-review literature is.

This can only mean that warmists review warmists' papers and reject those with a contrary view thus preserving the consensus. This is not new. When *Geophysical Research Letters* showed signs of wandering off the consensus road, Tom Wigley suggested that they: "get the goods on its editor Jim Saiers and go to his masters at the American Geophysical Union and get him ousted."

In 1990, Phil Jones published a paper comparing rural and urban temperature measurement stations in Russia, China and Australia. A very narrow interpretation showed little difference between the stations. Phil Jones boasted to Michael Mann: "Recently rejected two papers (one for JGR and for GRL) from people saying CRU has it wrong over Siberia. Went to town in both reviews, hopefully successfully."

It appears that Jones is so precious that he cannot take criticism from his peers and prevents peer-reviewed publication by his critics. After the Climategate emails were public, Jones posted the list of Russian stations used and the claim from Russia came that Jones had cherry-picked his data.

And, just to make sure that publicly funded scientists do not have to release information that underpins their predictions, Phil Jones commits conspiracy in an email to Michael Mann: "The two MMs [Steve McIntyre and Ross McKitrick] have been after the CRU data for years. If they ever hear there is a Freedom of Information Act now in the UK, I think I'll delete the file rather than send to anyone…"

This is consistent with Jones' email to Warwick Hughes: "Why should I make my data available to you when your only objective is to find something wrong with it?"

The Climate Research Unit computer programmer Harry Harris kept extensive notes on the defects he had found in the data and computer programs that the Climate Research Unit uses in its compilation of its global mean surface temperature dataset. These 15,000 lines of notes (Harry_Read_Me.txt) give the game away. He lamented:

[The] hopeless state of their [CRU] database. No uniform data integrity, it's just a catalogue of issues that continues to grow as they're found...I am very sorry to report that the rest of the databases seem to be in nearly as poor a state as Australia was. There are hundreds if not thousands of pairs of dummy stations, one with no WMO and one with, usually overlapping and with the same station name and very similar coordinates. I know it could be old and new stations, but why such large overlaps if that's the case? Aarrggghhh! There truly is no end in sight.

And:

This whole project is SUCH A MESS. No wonder I needed therapy!

And:

I am seriously close to giving up, again. The history of this is so complex that I can't get far enough into before head hurts and I have to stop. Each parameter has a tortuous history of manual and semi-automated interventions that I simply cannot just go back to early versions and run the updateprog. I could be throwing away all kinds of corrections – to lat/lons, to WMSs (yes!), and more. So what the hell can I do about all these duplicate stations?

No wonder Jones did not want to release anything in response to Freedom of Information requests. It would have shown that the data was just not adequate to create a model. If the data is not bulletproof, then the theory of human-induced global warming can only be questionable.

The mess is not restricted to the Climate Research Unit of the University of East Anglia. Phil Jones acknowledges that the CRU data mirrors the NOAA data:

Almost all the data we have in the CRU archive is exactly the same as in the GHCN archive used by the NOAA National Climatic Data Center.

To make matters even more exciting, the Ria Novosti agency reported that the Moscow-based Institute of Economic Analysis

(IEA) issued a report claiming that the Hadley Centre for Climate Change had probably tampered with the Russian Climate Data:

> The IEA believes that Russian Meteorological station data did not substantiate the anthropogenic global-warming theory. Analysts say Russian meteorological stations cover most of the country's territory and that the Hadley Center had used data submitted by only 25% of such stations in its reports. The Russian station count dropped from 476 to 121 so over 40% of Russian territory was not included in global temperature calculations for some other reasons rather than the lack of meteorological stations and observations.

The stations not listed often show no substantial warming. The Hadley Climate Research Unit temperature database (HadCRUT) includes stations with incomplete data (highlighting the apparent warming) rather than stations with uninterrupted observations. The Russians concluded that the HadCRUT data included the incomplete findings of meteorological stations far more often than those providing complete observations. Stations used are located in populations centres influenced by urban heat that created 0.64°C more warming than if 100% of the raw data was used. Considering that Russia is 11.5% of the Earth's land surface area, this significantly altered the global surface temperature. The same bias was used for polar Canada.

If the evidence for human-induced global warming is so strong, then there is absolutely no need to resort to the tactics exposed in the Climategate emails. There is only one conclusion. The keepers of the data upon which the IPCC comes to conclusions cannot be trusted. If they were comfortable with the accuracy of their methods and data, they would be proud to share it amongst the scientific community. Instead, they treat public information as their own and are quite prepared to destroy historical data rather than have it checked. If the data of the CRU cannot be checked and their computer codes for analysing data they admit are very weak and the HadCRUT data is unreliable, then anything written by the IPCC cannot be trusted because it cannot be validated.

The scientists in the Climategate email scandal act as fraudulent political activists. They have no respect for data, science and honesty. They have been involved in one of the biggest scientific frauds in the history of science. They show that they try to stifle data and debate. It does not look as if the climate industry is on a mission to save the world, it looks more like a game of power. They will stop at nothing to prevent their work being evaluated, investigated or validated. They are the leading lights in the IPCC and it is their work and influence that drives the IPCC.

Governments have relied on the UN's IPCC for dispassionate transparent information and all they have received is fraud. Yet governments and the climate industry quote the IPCC as the authority underpinning their beliefs.

Some of the so-called IPCC experts leave a lot to be desired. The IPCC chair Rajendra Pachauri explained how IPCC authors are selected:

> This is a very careful process of selection...These are people who have been chosen on the basis of their track record, on their record of publications, on the research that they have done...They are people who are at the top of their profession as far as research is concerned in a particular aspect of climate change...you can't think of a better set of qualified people than what we have in the IPCC.

Many IPCC authors are activists who clearly do not fulfil these criteria. If journalists peered deeply into the IPCC they would find Greenpeace. There are a huge number of clues around that don't pass the smell test. Why did Rajendra Pachauri write the foreword to a Greenpeace document that focussed on New Zealand? Why was Richard Klein, a Greenpeace campaigner, appointed as a lead author of the IPCC at the age of 25? Was this because of a long and illustrious career in science? Surely not. Bill Hare spent decades with Greenpeace, was their chief climate negotiator in 2007 and was nominated and chosen to fill senior IPCC roles as a lead author, expert reviewer for two of the three sections of the report and was one of the select 40 people in the core writing team for the Synthesis Report. He is lead author in the forthcoming 2013 IPCC report and the world waits breathlessly for his dispassionate consideration of the breadth and body of science. Another Greenpeace heavy since at least 2001, Malte Meinshausen, was a contributing author of three chapters of the 2007 IPCC Report. Considering that he was only awarded his doctorate in 2005, he is hardly a seasoned scientist who can bring disciplined detached science to the IPCC. In fact, the IPCC 2007 Report cites research papers by Hare and Meinshausen as if they are senior scientists. Even one of their graphs has been uncritically reproduced.

In another blow to an organisation set up to offer impartial science-based advice to politicians the IPCC was forced to admit that it had exaggerated the threats of global warming. The end result of IPCC tactics has been a change in government policy. In 2011 in the UK, the government shifted £8.6 billion from the poor to the middle classes and the wealthy as subsidies for wind and solar power generated at home and fed at four times the normal price into the electricity grid. The UK government has agreed to pay solar power generators £0.239 per watt of solar electricity yet the normal cost of electricity is £0.085 per watt. This gold rush triggered the spoiling of much of the English countryside with solar farms.

The 2007 IPCC report had David Karoly as the review editor for Chapter 9. He must have read comments submitted by Vincent Gray, a meteorologist of 60 years standing but he did not even show Gray the courtesy of replying. This is not how the peer review process should operate. Chapter 9 is the sole chapter claiming that there is human-induced global warming and attributing it to human-produced carbon dioxide. It built on Chapter 12 of the 2001 Report (in which David Karoly was lead author). Chapter 9 was produced by a tightly knit club of people dominated by computer modellers who were neither independent nor objective. In the most important chapter in the 2007 report, the IPCC review process was corrupted and commonly completely bypassed. Almost two thirds of references cited in Chapter 9 were written or co-written by chapter authors (dominated by the Climategate stars). The IPCC chair Rajendra Pachauri claimed that 100% of the science is peer-reviewed yet an independent audit found that 5,587 references were not peer-reviewed; they included newspaper articles, hikers' anecdotes and political activists' campaign material.

What is omitted from the IPCC reports is even more interesting. Why has the IPCC omitted huge bodies of geology, astronomy, solar physics and cosmic ray studies that present a different story? Why did the IPCC omit 90,000 measurements of atmospheric carbon dioxide made over the last 180 years by a validated chemical method? Why did the IPCC not state that ice core measurements show that in the past atmospheric carbon dioxide rise has been a consequence of temperature rise, not as a cause?

Even the scientific journals have distorted the peer-review process. Greenpeace co-founder and ecologist Patrick Moore has slammed the journal *Nature* for uncritically publishing activist science on possible mass extinctions. He has distanced himself from Greenpeace who he claims is a political activist lobby and not an environmental group. Sari Kovats was selected as a lead author by the IPCC. In 1994, she was an unknown. By 2007, she was the lead author on the health chapter and was a contributing author for three other chapters (Ch. 1: Assessment of observed changes and responses in natural and managed systems; Ch. 2: Coastal systems in low-lying areas; Ch. 12: Europe). Some three years later, she received her doctorate, the degree usually considered the basic meal ticket for commencing a research career. She fulfils none of Pachauri's criteria and certainly has not been in the game long enough to be considered an expert in anything.

When we hear a climate industry activist claim that science supporting human-induced climate change is published in the peer-reviewed literature and that the views of contrarians are not published, then we can only conclude that these activists are knowingly supporting an agenda. The Climategate emails are there for all to see. This is exacerbated by the UK's Climate Research Unit cherry picking data to show that there had been an extraordinary increase in global temperatures in the late 20th century and withholding data from researchers who may be contrarians.

Al Gore's film *An Inconvenient Truth* was perfect. It allowed him to set the scene for his carbon trading activities by frightening us witless. In 2007 in the UK, the Department of Education and Skills announced that all secondary schools were to be provided with a climate change pack that included this film. Many parents were incensed that schools were being used for propaganda because Gore's film had been shown to be scientifically wrong. Not arguably wrong, but hopelessly wrong and fraudulent. Legal action was taken by a parent against the Secretary for Education to withdraw the film from schools. In 2007, Justice Burton of the High Court ruled that Gore was on a crusade, that the science of the film had been used to support a political program and that the film contained fundamental scientific errors. Charles Monnett, the US federal biologist whose headline-grabbing claim that global warming was drowning polar bears (used as a key piece of evidence in Al Gore's film) has now been suspended while authorities investigate allegations of scientific misconduct, reportedly in the polar bear study that made him famous.

Despite this, in October 2010 it was decided that Gore's film should be included in the new national English curriculum in Australia as part of a bid to teach students about "environmental sustainability" in all subjects. The film is scientifically wrong, it is political propaganda and has nothing to do with undefined "environmental sustainability". Imagine the horror if a pupil learns that the climate is OK and that Gore's propaganda is science fiction. Activists, advocates, the media and teachers will lose respect and their power and the media will have to dream up something else to keep the masses frightened. It is healthier for the country to teach the basics of language, reasoning, mathematics, history and science and, by then, pupils will have the skills to critically analyse "environmental sustainability" with clarity of thought. They would then have the basic tools to solve environmental problems.

Out there in media land, we often hear that 4,000 scientists reviewed the IPCC Reports. The IPCC chair Rajendra Pachauri and former Australian PM Kevin Rudd repeat this chant over and over again and the media buy it. Yet an investigation shows that the IPCC's latest report had only five reviewers and there is nothing to suggest that they were scientists. This information was only released by the IPCC when it feared the use of Freedom of Information laws to obtain the names of the reviewers.

Why does the IPCC change the text of scientists to give the opposite meaning? For example, the original Chapter 8 of the 2007 IPCC report read:

> Finally, we have come to the most difficult question of all. 'When will the detection and unambiguous attribution of human-induced climate change occur?'

In the light of the very large signal and noise uncertainties discussed in the Chapter, it is not surprising that the best answer to this question is, "We do not know". Although it is a loaded question that presupposes that human emissions of carbon dioxide actually change climate, it is clear that the IPCC scientists cannot show that human activities create global warming. This was changed by the editor Ben Santer to make sure that the Chapter's findings were in accord with the Summary for Policymakers that was written by activists, bureaucrats and politicians. Scientists do not write the Summary for Policymakers yet this Summary is allegedly based on the science in following chapters. Santer changed the scientists' statement to: "The body of statistical evidence in Chapter 8, when examined in the context of our physical understanding of the climate system, now points toward a discernible human influence on global climate".

This fraud did not go unnoticed. Frederick Sutz, President Emeritus of Rockefeller University, one of the most distinguished scientists in the world, wrote in the *Wall Street Journal:*

> In my more than 60 years as a member of the American scientific community, including service as president of both the National Academy of Sciences and the American Physical Society, I have never witnessed a more disturbing corruption of the peer-review process than the events that led to this IPCC report.

This is fraud is no surprise. Santer has form. In the IPCC 2005 report (Chapter 8), Santer presented balloon evidence to show that the upper atmosphere was warming between 1965 and 1987 and that this was a fingerprint of global warming. He did not use the complete data set that showed cooling (1958-1965), warming (1965-1987) and cooling (1987-1995). The complete data set showed no warming trend whatsoever. Santer had only chosen the part of the graph that supported his ideology.

The UN's scare story of climate refugees came to a sticky end. In 2005, the UN's environment program predicted that by 2010, there would be 50 million climate refugees. This clearly did not happen and the map showing where the climate refugees were supposed to come from was removed from the web site. Screen shots of the map clearly show the title: "Fifty million refugees by 2010." The map vanished and is replaced by an error message that reads:

> Dear visitor, it seems that the map you are navigating by is maybe not fully up-to-date, or that it might have an error in it, or is it that your GPS is not loaded with the correct data?

In my view this is a lie. The map was removed because it showed that the scary prediction was wrong. Not only are we missing 50 million climate refugees, we are also missing the map showing where they would come from.

The Great Barrier Reef is meant to be the canary in the cage for human-induced global warming. But it just does not do what its warming industry masters want it to do. Ove Hoegh-Guldberg is the self-appointed champion of the Great Barrier Reef. He also seems to have a love affair with Greenpeace. And so he should. They put bread and wine on his table and he has been cashing cheques from activist groups for 17 years. There is nothing necessarily wrong with this but his funding from the most influential activist groups in the world needs to be declared. Furthermore, it is very difficult to believe that he is a disinterested party who can write scrupulously objective reports for the IPCC. His role on the IPCC as co-ordinating lead author shows that he is just another IPCC scientist compromised by Greenpeace. Nine of the chapters in the 2007 IPCC report had input from Hoegh-Guldberg or used his work. A report for Greenpeace on coral reef bleaching in French Polynesia – a cheque. Another report for Greenpeace on central Pacific reefs – another cheque. Another coral bleaching report for Greenpeace – yet another cheque. A Great Barrier Reef report for the World Wildlife Foundation – another cheque. And on it goes, cheque after cheque after cheque. He was an expert witness for an Australian government tribunal and he listed his 10 major research papers. Four were published by Greenpeace and a fifth by the World Wildlife Foundation. Five documents for activist groups can hardly be called major research papers. He has been asked many times to provide evidence that human production of carbon dioxide causes global warming. He has not. Yet during an ABC TV *Lateline* interview (29th October 2010), he did not state that he could provide such evidence and his writings and those of the IPCC contradict his media statements. If Ove was in the corporate world, failure to fully disclose his financial interests would result in him having a few years on a bread and water diet with an hour of exercise a day.

As soon as the Climategate scam broke, rather than address the issues, the University of East Anglia appointed a public relations person to handle the issue. Various fellow travellers joined in a smear campaign of those who did not sign a petition of support for the perpetrators of the scam. The whole grubby affair is laid out on Steve McIntyre's Climate Audit blog. If the University were really concerned about its reputation, it would have immediately got to the bottom of the matter rather than introducing another layer of spin.

There have been long and painful attempts to get the CRUTEM temperature data from the University of East Anglia. If the University had nothing to hide it would have proudly released the data for all to see. Freedom of Information requests by Steve McIntyre were rejected by the University because the temperature information may have had an effect on: "international relations, defence, national security and public safety", "intellectual property rights" or "the interests of the person who provided the information".

The first two reasons are nonsense. One wonders how temperature data can affect national security and public safety. For the University to claim intellectual property rights on information paid for by the taxpayer and used to frighten the taxpayer witless is beyond the pale. The third however is quite correct. Phil Jones of the Climate Research Institute would not want any data released because it would show poor housekeeping, gaps in the data, "adjusted data" and the flawed mathematical foundation of the computer models.

To treat basic climate measurements as a national secret shows that the whole business is on the nose. Science does not produce conclusions from secret data that cannot be replicated or reviewed. To make matters worse, Phil Jones admitted in a BBC interview that there had been no statistically significant warming for the last 15 years. This shows that Jones, the Climate Research Unit of the University of East Anglia and the whole climate industry need to come clean and show the raw data that on one day shows dangerous human-induced global warming and on another day shows that there is really nothing to worry about.

A number of Freedom of Information applications to the University of East Anglia were rejected and appeals to the University of East Anglia fell on deaf ears. Jonathon Jones of Oxford University appealed to the UK Information Commissioner against the decision to withhold information. The Commissioner ordered that the University of East Anglia provide the data that they had previously provided to Georgia Tech University. This means that the original data used by the University of East Anglia's Climate Research Unit (which underpinned IPCC reports) can finally be checked after years of bitter dispute. The Climate Research Unit has now released most of the raw data from 5,113 weather stations around the world it had used to piece together its highly contested land temperature data set. Not all the data was released; data from 10 Polish stations is still missing. Although the adjusted gridded data has been available for some time, this is the first time that the raw data has seen the light of day.

The University of East Anglia took James Delingpole, the *Spectator* writer and very popular blogger for UK's *Daily Telegraph*, to the independent Press Complaints Commission. There is no doubt that Delingpole calls it as it is. For example:

> The FOI-breaching, email-deleting, scientific-method-abusing Phil Jones of the University of East Anglia, for example, has granted tame interviews in Nature magazine and The Times presenting himself as a man far more sinned against than sinning.

And:

> Obviously, if Sir Edward Acton wants me to go into a bit more detail about the grotesque inadequacies of the Climate Research Unit, why the University of East Anglia has become a standing joke, how the Climategate emails showed the scientists at the very heart of the IPCC to be untrustworthy, unreliable and entirely unfit to write the kind of reports on which governments around the world make their economic and environmental decisions, or why it is that Professor Phil Jones is more likely to find himself remembered for the Climategate scandal than he is to find himself mentioned in the same breath as Einstein, Newton or Watson and

Crick, I should be more than happy to do so.

The Press Complaints Commission report of 9th April 2011 was a crushing repudiation of the University's attempt to manage dissent and stop free speech. The report stated:

> The Press Complaints Commission was satisfied that readers would be aware of the context of the columnist's robust views – clearly recognisable as his subjective opinion - that the scientists were "untrustworthy, unreliable and entirely unfit to write the kind of reports on which governments around the world make their economic and environmental decisions", and that their work was "shoddy" and "mendacious". In the circumstances, it did not consider that there had been a breach of Clause 1 (Accuracy) of the Code.

Tree-ring data from Russia are critical for the understanding of recent climate changes and comparing them to the present changes. Freedom of Information requests to the University of East Anglia for the data that supports the tree-ring work were to no avail. Not only was the data request refused, there was a refusal to identify sites where the data was collected. Validation of data is vital for science. If one of the Climategate universities refuses to release publicly-funded data for scrutiny then they don't pass the smell test. One would think that scientists who claim to be concerned about the future of the planet should be able to support their views from even the most rigorous scrutiny. They won't. This suggests that they can't. This suggests that Climategate is just the tip of the iceberg. This is the big story for an investigative journalist. Western countries have already invested billions in climate research which may be the most wasteful fraud to date in the history of humanity.

To construct the use of so-called average global temperatures is full of errors and uncertainties. The Earth's climate is not homogeneous but is in fact a highly dynamic chaotic system both in terms of geography and time and hence the term "average global temperature" is meaningless. To average the coldest winters in Europe and South America with the hottest summer in Russia to make a pronouncement on global temperature is ludicrous.

Averaging temperatures in Canada is meaningless. What does the average between the relatively warm wet Vancouver winter and the very cold winter in Montreal really mean? In fact, the so-called average global temperatures are partly based on large areas of the world that have no weather stations, more than 60% of these having been decommissioned since 1990. Many others are not maintained or are in locations subject to urban heat island effects. The gaps are filled by the guesswork of the same type of computer models used by the UK Meteorological Office to predict mild winters when they were in fact perishingly cold. Some of our Climategate heroes have got themselves in an even greater spot of bother. Phil Jones created a global data set for temperature and was not aware that much of the data from China for his computer models had been fabricated.

Just before the 2009 exposure of the Climategate fraud, the IPCC chair Rajendra Pachauri urged us to change unsustainable life styles, eat less meat, not drink iced water, pay extra for hotels with air conditioning and avoid the "irrational" decision to fly. As IPCC boss, he flies all over the world and stays in the best of hotels. As a cricket-lover, he flies to watch games in many parts of the world. Despite claiming that he has gained no benefit from his IPCC activities, he lives in one of Delhi's most expensive areas, has a web of business interests, directorships and advisory positions on matters climate and energy. These had not been declared but had been dug out by investigative journalists. Pachauri is a director of The Energy Research Institute (TERI) in New Delhi which claims to be a non-profit group with the brief to "work towards global sustainable development, creating innovative solutions for a better tomorrow". Land given by a developer for "institutional or public or semi-public purpose" was converted to a golf course in a parched area of India. This sustainable development uses 1.2 million litres of water a day and sells club membership to rich Indians for $500. The Australian Department of Climate Change and Energy Efficiency provided two grants ($48,340 and $45,000) to Pachauri's TERI so I guess the Australian taxpayer owns at least one of the fairways on the golf course.

In Pachauri's IPCC 4th Assessment Report, it was claimed that by 2020, climate change would cause agriculture in Africa to fall by 50% because of lack of rain. And the source of this shocking revelation? A report of a Canadian advocacy group written by an obscure Moroccan academic who specialises in carbon trading. The claim that potential reduced rainfall caused by climate change threatens the survival of 40% of the Amazonian rainforest originated from a World Wildlife Fund document dealing with logging and forest fires, not climate change.

The IPCC 2007 report claimed a very high probability that Himalayan glaciers would disappear by 2035 and that the people of the Indian sub-continent would die of thirst. There was a simple problem: it was not true. The claim derived from a 1999 Syed Hasnain interview with *Down to Earth*, an Indian environmental magazine, and a similar interview given to *New Scientist*. Why didn't *New Scientist* show some editorial scrutiny? Georg Kaser, an Austrian glaciologist, stated it was rubbish as did Indian glaciologist Vijay Raina. Rajendra Pachauri defended Hasnain's claims by accusing Raina of "voodoo science". It was shown that Hasnain was an employee of TERI's new glaciology unit that had been created to suck up a $500,000 research grant to study Himalayan glacier melting. Pachauri found himself under pressure to resign and limply stated: "I don't think it takes anything away from the overwhelming scientific evidence of what's happening with the climate on this Earth" and that it was: "human error".

Less charitable folk would call this fraud. Hasnain pathetically claimed that he was misquoted.

Exaggeration is the strong hand of the IPCC. In their Summary for Policymakers, the urgency became greater with time, probably because fewer and fewer people were influenced by the IPCC's catastrophism. In the First Assessment Report (1990) we learned that: "The observed [20th century temperature] increase could be largely due…to natural variability."

Now this is not what the scaremongers wanted the public to hear, so the environmental activists changed the science to: "The balance of evidence suggests a discernable human influence on climate " (Second Assessment Report 1996)," then to: "There is new and stronger evidence that most of the warming observed over the last 50 years it attributable to human activities" (Third Assessment Report 2001) and finally in 2007 to: "Most of the observed increase in globally-averaged temperature since the mid-twentieth century is very likely [=90 per cent probable] due to the observed increase in anthropogenic greenhouse gas concentration."

We await the Fifth IPCC Report where we will learn that we have already died from human-induced global warming.

It was not possible for these self-appointed guardians of the truth to monitor every scientific paper related to climate. Over 900 peer-reviewed papers and numerous scientific books were published in recent times in a great diversity of scientific disciplines contradicting the human-induced global warming mantra. As these papers were not written by the climate consensus club, have a contrary view, and escaped the censorship of our "climate scientists", it will be interesting to examine the Fifth IPCC Report to see how many of these papers are actually acknowledged. The Fifth IPCC report will be exposed to intense scrutiny, a normal process of science. Research supporting dangerous global warming has been a gravy train offering easy fame and fortune. The reward for critics has been personal denigration and defamation.

Snow, ice, floods and cyclones

On March 20th 2000, the British newspaper *The Independent* reported David Viver of the UK's Climate Research Unit warning that within a few years snowfall will become: "a very rare and exciting event" and that: "children just aren't going to know what snow is".

Similarly, David Parker at the UK's Hadley Centre for Climate Prediction and Research said that eventually British children could have only a virtual experience of snow via movies and the internet. The last three winters in the UK were forecast by the UK Met Office to be mild and snowless. Instead, there was brutal cold and snow in the UK. The UK was shut down for a few days. The public were not convinced that these blizzards were due to global warming. Those claiming to forecast climate in 100 years time could not even use their massive supercomputers to get the weather right for the next month. They now want bigger and better supercomputers, which will only give them the wrong answer more quickly. Some UK parliamentarians called for an investigation as to whether the UK's Met Office's reliance on their ideology and carbon dioxide models had biased their predictions. They have realised that there is no evidence that carbon dioxide drives global temperatures and that human-induced global warming is just a theory derived from the manipulation of models in supercomputers that few understand and that have little bearing on reality.

The UK government plans to spend £1 billion on carbon capture and burial schemes and untold billions on wind power subsidies, but wind provided almost zero power when needed in the recent exceptionally cold winters. In places, the wind towers actually consumed electricity to protect them from frost damage. UK residents would have been better off had all of that money been spent on reliable power sources and snow-proof infrastructure. The UK government also spends £200 million a year on the Met Office who have made failures in long-term prediction into an art form. Just before the Copenhagen climate summit, the UK Met Office tried to scare the public by suggesting that the mean world temperature for 2010: "is expected to be 14.58°C, the warmest on record."

This was really scary as the predicted temperature would rocket an astronomical 0.58° above their 1961-1990 average. Such predictions are made using the poor order of accuracy of old stations, data from closed stations and massaged measurements. The timing of the announcement was clearly made for political reasons and not to inform the public. This prediction was made while the UK was in the grip of snowstorms and freezing weather. The public did not buy it. It later emerged that the data from the University of East Anglia's Climate Research Unit and the UK Met Office's Hadley Centre showed that 2010 had been cooler than 2005 and 1998 and equal to 2003. It emerged that, for the purposes of the press release, the data had been significantly altered and the 2010 data that was adjusted upwards.

In late 2010 while climate delegates at the climate summit in the resort paradise of Cancún (Mexico) were trying to save the world, the UK Met Office was at it again by stating: "Our experimental decadal forecast confirms previous indications that about half the years 2010-2019 will be warmer than the warmest year observed so far – 1998."

What was not said was that both 1998 and 2010 were big El Niño years and that during El Niño years, the warming of the Pacific Ocean affects the world. The UK Met Office states that warming since the 1850s has been 0.8°C but they don't tell us which part of this is natural and which part may be man-made. In a place renowned for its miserable climate, I am sure that the average Englishman would enjoy an extra 0.8°C (whatever its origin). A more reliable rule of thumb for the UK Met Office for their long-term predictions would be to prepare for the opposite or even look out the window. Private UK weather forecasters like Piers Corbyn predicted the bitterly cold winters for 2009 and 2010 on virtually no budget. He got it right. Corbyn does not use supercomputers but uses the Sun. The UK Met Office ignores the Sun.

In the USA from 1993 to 2006, 2.2 people per million died in events of extreme cold compared to 1.3 in events of extreme heat so if we want to fear something, fear the cold and not the heat, as people have done for thousands of years.

We have daily weather predictions. These predictions have an extraordinary degree of accuracy and we really only notice those that are significantly different from the prediction. These are short-term predictions. Longer-term seasonal predictions are far more inaccurate and predictions of weather and climate in 100 years time are fanciful. On 24th August 2010, the Australian Bureau of Meteorology predicted a warm dry spring. Instead, it was wet and cold. For south-western Western Australia, the prediction was for a wetter than normal spring. It was not. It was drier than normal. For the rest of Australia, rather than being the predicted dry and warm it was wet, wet and wet with some places recording the highest rainfall on record. It was also predicted that spring was going to be "hot across the north". Instead, it was cool. The same methods used to predict the weather are used to predict climate 50 to 100 years in advance. With weather and long-term climate predictions, we must be very cautious and should have little confidence in long-term predictions.

Australian governments are spending at least $800 million annually on weather monitoring, meteorology and climate research. Major droughts and empty dams were predicted for 2010-2011, flood mitigation projects did not take place and there seemed to be total bewilderment when the 2010-2011 summer floods occurred. Some scientists had warned that changes taking place in the Pacific Ocean would lead to a return of flood years like 1974. Private weather forecasters were aware that the Southern Oscillation Index was falling well before the big wet of the 2010-2011 summer suggesting that a La Niña event was on the way. In a major report, the Queensland Government's climate change folk did not even mention the possibility of floods in a major report. Floods occurred soon after the report's release. Of course, the floods we later heard were due to global warming. Environmentalists and alarmist scientists have reinvented global warming and now attribute all weather to global warming whether there be warmth, cold, drought or flood. They now call it "climate disruption". Whatever the label, they were unable to forecast the massive floods that devastated much of eastern Australia. That $800 million would have been better spent on water storage and flood-proofing roads, bridges and airports.

The Australian Bureau of Meteorology states that trends in tropical cyclone activity in the Australian region show that the total number of cyclones has decreased in recent years.

Tropical cyclone numbers in the Australian region are influenced by the El Niño-Southern Oscillation phenomenon and the decrease may be associated with an increased frequency of El Niño events. On 1st February 2011, the deputy leader of a political party (the Greens Party) stated that tropical cyclone Yasi is a: "tragedy of climate change" and "Scientists have been saying that we are going to experience more intense weather events, that their intensity is going to increase their frequency."

This is contrary to what has been stated yet ABC News decided to run this nonsense uncritically as a news story without telling the audience that Queensland had far worse cyclones and floods a century ago. They did not ask the Green Party politician simple questions about the past. There is a good historical record of floods of the Brisbane River since 1824. Many of these floods were fatal and there have been larger floods than those in 2010 and 2011. Maybe the ABC was happy to broadcast nonsense in accord with their political advocacy? The Bureau of Meteorology web site would be a good start for the ABC if they had wanted to present news rather than sensationalist environmental activism.

The Bureau of Meteorology showed that on average 4.7 tropical cyclones per year affect the Queensland Tropical Cyclone Warming Centre's area of responsibility. There is a strong relationship between eastern Australian tropical cyclone impacts and El Niño-Southern

Oscillation phenomena with almost twice as many impacts during La Niña compared with El Niño periods. There have been 207 known impacts of tropical cyclones along the east coast of Australia since 1858. Major east coast tropical cyclone impacts were in 1890 (Cardwell), 1893 (Brisbane), 1898 (northern NSW), 1899 (Bathurst Bay), 1918 (Innisfail), 1918 (Mackay), 1927 (Cairns), 1934 (Port Douglas), 1949 (Rockhampton), 1954 (Gold Coast), 1967 (southern Queensland), 1970 (Whitsunday Islands), 1971 (Townsville), 1974 (Brisbane) and 2006 (Innisfail). The Queensland section of the Gulf of Carpentaria has had disastrous cyclones in 1887, 1923, 1936, 1948 and 1976. Conclusion! Don't believe what you read in newspapers, what you see on television or hear on the news.

Cyclone Yasi's central pressure was 922 hPa. It was not as low as the 914 hPa of Cyclone Mahina which killed more than 400 Queenslanders in 1899. Over time, Queensland averages two cyclones a year, each of which kills more people than in the Queensland 2010 natural disasters. We have had far costlier disasters than the 2011 floods including Cyclone Tracey (Darwin, 1974, 71 killed) and the 1989 Newcastle earthquake. What we are seeing is that we are now better prepared for natural disasters. What we don't see is any proof that Cyclone Yasi resulted from human activity emitting carbon dioxide into the atmosphere. If this was the case, we should send the bill to China and the US.

As for 2010 being a year of global cyclones, the NASA GISS record has a different story. In 2010, there were 66 tropical cyclones globally. This was the fewest since reliable records have been kept. There were 46 tropical cyclones in the Northern Hemisphere (fewest since 1977). Hardly a month goes by when some weather record is not broken. This is largely because of better communication systems rather than changes in global weather and climate.

The IPCC's Fifth Report is in the planning stage. We wonder if fraud again is planned. The list of Australians involved in the IPCC Fifth Report very much looks as if it will be more exaggeration, more misleading and deceptive science, more "adjusted" data, more secret computer codes, more dodgy data from unreliable sources, more omissions of contrary science and more fraud. We wonder if governments will again blindly accept the IPCC's report compiled by those with previous form.

What have those in the climate industry delivered to society? They have promoted a shallow, implausible and inconsistent hypothesis. It has refused to address well-founded objections from eminent scientists and, instead, reverted to attacking the character of those who question its hypothesis. It has attempted to destroy the careers of scientists who differ and have accepted a hypothesis as dogma. This has taken place concurrently with the trashing of the protocols of the scientific method, the takeover of once respectable scientific journals and corruption of the peer review process. The climate industry has changed the nature of "climate science" by stating until they are blue in the face that there is no debate, that debate has finished, that there is "consensus" and that the "science is settled". They have distorted raw data, refused to make raw data and their methodology available for validation and refused to acknowledge inconsistencies between adjusted surface temperature data and satellite and radiosonde data. They promote, where convenient, weather events as proof of human-induced global warming and, where convenient, conversely argue that weather and climate are not the same.

Maybe, as part of national due diligence, independent scientists should be funded to recalculate and re-evaluate every claim made by climate catastrophists. After all, every claim made whether it be with regard to sea level, ocean surface temperature, ice melting or atmospheric temperature change was shown to be exaggerated and there is no evidence to suggest that over-icing the cake will stop.

Just because the data is cooked does not mean that the planet is cooking. No wonder many say that the climate industry is promoting the greatest scientific fraud ever seen on this planet.

Fellow travellers

Most journalists have enthusiastically embraced the possibility of a catastrophe derived from human emissions of carbon dioxide. It sells stories. Most journalists know no science and are therefore incapable of having an informed opinion on human-induced global warming. Many journalists have given away their independence and impartiality by supporting political advocacy and shutting out the informed opinions of eminent scientists who challenge the climate industry.

Herald Sun columnist Jill Singer enthusiastically supported a tried and proven method of silencing debate:

> I ...propose another stunt for climate sceptics - put your strong views to the test by exposing yourselves to high concentrations of carbon dioxide or some other colourless, odourless gas - say, carbon monoxide. You wouldn't see or smell anything. Nor would your anti-science nonsense be heard of again. How very refreshing.

Richard Glover, ABC 702 Sydney, suggested:

> Surely it's time for climate change deniers to have their opinions forcibly tattooed on their bodies...Just something along the arm.

Tattooing and gassing sceptics? Does a history analogy perhaps suggest itself.

Elizabeth Farrelly in *The Sydney Morning Herald* yearned for the virtue of a communist dictatorship, which, unlike Australia, needs not to pander to the 80% of the population that has an IQ of just 80 and resists virtuous causes such as global warming. Farrelly presumptuously assumes that she is in the top 20% of the intellectual elite and hence is born to rule dictatorially. She did not seem to be aware that an undemocratic country like China has no truck with the human-induced global warming nonsense.

Former Green Party political candidate Clive Hamilton stated:

> The implications of 3°C, let alone 4°C or 5°C, are so horrible that we look to any possible scenario to head it off, including the

93

canvassing of "emergency" responses such as the suspension of democratic processes.

Hamilton clearly demonstrates by such a statement that he knows no history or science and, while blessed with such ignorance, is advocating despotism. He could only make such statements in a free country and yet wants to withdraw freedoms. Furthermore, he suggests that a warmer climate is horrible. Most of the world longs for warmth.

And, just to make sure that we use the processes of despots, David Shearman (IPCC assessor and Secretary for Doctors for the Environment Australia) suggested: "(green) government in the future will be based upon....a supreme office of the biosphere" comprising "specially trained philosopher/ecologists" who "will either rule themselves or advise an authoritarian government of policies based on their ecological training and philosophical sensitivities. These guardians will be specially trained for the task".

One does not have to go back far in history to see certain guardians trained to maintain a philosophical position. My Chinese students who survived the Cultural Revolution have experienced Shearman's philosophy. When Lord Monckton talks of eco-fascism, he might not be wrong. When President Klaus asks "What is at stake: climate or freedom?", he has hit the nail on the head.

In recent times, anyone who actually dares to question whether there may be catastrophic global warming is labeled a climate denier, an implicit reference to Holocaust denial. Nobody denies climate change, just the same as nobody denies changes in weather. Again, an historical analogy suggests itself. There are many countries in the world where if you just happen to ask a few too many questions, you disappear and end up as fish food. Is this really what the strident environmental activists want for democratic countries?

The planet is cooling in the 21st century and the scientific credibility of warmists has collapsed. Again. If the planet is cooling and the warmists blindly state that it is warming, one wonders who the denialists really are?

Oh what freezing warming we are currently enjoying!

2

SCIENCE

The process of science

Science is married to evidence. This evidence is collected by observation, measurement and experiment. Computer models do not constitute evidence. In science, the debate is about whether the process of evidence collection was valid, what the errors were, what assumptions were made and whether such assumptions were valid. A body of evidence is interpreted and explained. This explanation is a scientific theory. In the law, some evidence may be inadmissible which is why some legal decisions seem bizarre. Not so in science. Some scientific ideas such as cold fusion are very quickly abandoned when the phenomenon can't be reproduced and validated. The real peer review process occurs after publication of a paper.

There was once a consensus about eugenics. The plan was to identify those that were feeble minded (and that variously included Jews, blacks and foreigners) and stop them from breeding by isolation in institutions or by sterilisation. George Bernard Shaw claimed that it was only eugenics that could save mankind. H.G. Wells spoke against: "ill-trained swarms of inferior citizens" and Theodore Roosevelt stated that: "Society has no business to permit degenerates to reproduce their kind."

Eugenics research was funded by the Carnegie Foundation and the Rockefeller Foundation. With some eminent citizens and respectable foundations supporting eugenics, how could it not be science? We now know it was a racist, anti-immigration social program masquerading as science.

97

Government ideology and an uncritical media have led to the promotion of false scientific concepts. One example was Trofirm Denisovich Lysenko, a self-promoting Russian peasant who invented a process called vernalisation. Seeds were moistened and chilled to enhance later growth of crops without fertiliser and minerals. It was claimed that the seeds passed on their characteristics to the next generation. Stalin wanted to increase agricultural production, vernalisation never had scientific scrutiny, Lysenko became the darling of the Soviet media and was portrayed as a genius and any opposition to his theories was destroyed. Lysenko's theories dominated Soviet biology. They did not work. Millions died in famines, hundreds of dissenting scientists were sent to the gulags or the firing squads and genetics was called the "bourgeois pseudoscience". History has a habit of repeating itself.

Scientific evidence must be reproducible. It matters not whether the evidence is from Chad, Colombia or China, the evidence must be able to be reproduced. Science transcends cultures, religion, gender and race because any validated scientific evidence is validated scientific evidence. The scientific theory will change with new validated evidence hence it is impossible in scientific discipline for the science to be settled. Science is anarchistic, has no consensus, bows to no authority and it does not matter what a scientific society, government or culture might decide, only reproducible validated evidence is important. Like any other area, science suffers from fads, fashions, fools and frauds, has its short-term leaders and can be cultish.

Scientific ideas can be tested. There may be one hundred sets of evidence in support of a scientific idea. It only takes one piece of validated evidence to the contrary and that idea must be rejected. The idea examined in this book is that human emissions of carbon dioxide drive global warming and that this warming may be irreversible and catastrophic. This is easily tested by comparing with the reproducible validated evidence from past climate changes. This is the coherence criterion of science. Any new idea needs to be in accord with validated evidence from other areas of science. Human-induced global warming is not in accord with geology. There are hundreds of examples that show that past events of global warming were not driven by carbon dioxide, that the planet has been far colder and warmer in past times and that the rates of temperature change have been far faster than any changes measured today. The past is a story of constant climate change and our present climate is the result of the past. The idea that human emissions of carbon dioxide that lead to catastrophic global warming is therefore invalid and to continue to promote such an idea is ignorance, fraud or perhaps both. No wonder those that call themselves "climate scientists" don't want to debate geologists. The "climate scientists" solution to this spot of bother is to ignore geology, astronomy and history and yet claim that they are involved in science. All this shows is they are involved in political and/or environmental advocacy.

All scientific evidence and the ideas created from this evidence have a degree of uncertainty. To make strong statements or predictions without declaring the uncertainties is not science. Geologists learn this very quickly. Testing of ideas hidden below with a drill hole is a very humbling experience. For scientists to declare certainty, to make predictions and to ignore large bodies of contrary science shows that personality weaknesses overwhelm the scientific method.

Every active scientist is constantly involved in scientific disputes because science is never settled, new data and ideas are continually being aired and any new idea is generally critical of previous work. That is the nature of science. Many of us have written papers with our critics and rarely do scientific disputes become personal. Some of us have been editors of major scientific journals, sat on editorial boards, refereed scientific papers, sat as experts on research council panels, examined doctoral theses, been called as expert witnesses in Court and are scientific authors.

A scientific paper or grant proposal can easily be buried by sending the work to a referee in another camp. Peer-review is only an editorial aid. It involves rejecting or accepting a paper and making suggestions and changes to the submitted paper. The number of reviewers that can be called on for a submitted paper is generally small. Peer-review is conducted at the discretion of editors and the peer-review process is very susceptible to the influence and bias of small groups promoting and protecting their own interests. Most reviewers are anonymous. This only serves to strengthen the influences and biases. Pre-publication review is important but the most important process occurs after review. This is when the scientific community at large can refute, check, confirm or expand on the ideas published. The climate clan is peer-reviewing and publishing their own grant-earning dogma and some of the scientific community is showing how weak this science is post-publication. Peer-review, far from being the golden standard, is used by the climate industry to approve or censor science according to its own agenda.

After a lifetime of science, I have seen lots of shoddy published science and failings in the peer-review system. For example, Jan-Hendrik Schön managed to publish a paper every eight days in major scientific journals between 2000 to 2002 on nanotechnology and single molecule behaviour. Every paper was peer-reviewed, every paper was bogus and yet Schön's great works were published by *Nature*, *Science*, *Applied Physics Letters* and *Physical Review*. Schön won a number of prestigious prizes for his published work. This fraud was later detected by a student, not by the peer-reviewers. Schön showed that if you have influential mainstream mates, fraud can easily be published in the best peer-reviewed scientific journals.

However, in the climate industry there are no parallels. There appears to have been a profession-wide decision that there can be nothing published that might threaten their view that there is human-induced global warming. This lack of scientific objectivity is such that one has to question whether any literature from the climate industry can be treated seriously for another 30 years. The climate industry has openly and flagrantly violated every aspect of the core ethos of science. Climate industry science is technically-based advocacy in the service of politics and is not dispassionate.

In the Wegman Report (which reviewed the Michael Mann hockey stick), the peer review process in the climate industry was questioned:

> One of the interesting questions associated with the "hockey stick controversy" are the relationships among authors and consequently how confident one can be in the peer review process. In particular, if there is a tight relationship among the authors and there are not a large number of individuals engaged in a particular topic area, then one may suspect that the peer review process does not fully vet papers before they are published.

And:

> Indeed a common practice among associate editors for scholarly journals is to look in the list of references for a submitted paper to see who else is writing in a given area and thus who might legitimately be called on to provide knowledgeable peer review. Of course if a given discipline area is small and the authors in the area are tightly coupled, then this process is likely to turn up very sympathetic referees. These referees may have co-authored other papers with a given author. They may believe they know that author's other writings well enough that errors can continue to propagate and indeed be reinforced.

In the case of the "hockey stick", there were only 43 in the climate industry who could have been gatekeeper. With such a small club, it is not surprising that shoddy science can be published.

In 2007, Ross McKitrick spent two years trying to publish a paper to refute a claim by the IPCC that real-world data confirmed models predicting man-made global warming. The IPCC claim was based on fabricated evidence and was shown to be fabricated using statistical analysis. Ultimately, the paper was accepted for publication but not in the climatology literature.

Developments in "climate science" that challenge the findings of the IPCC continue. A 2011 paper by one of the world's most eminent meteorologists, Richard Lindzen, addressed concerns raised by critics of a 2009 paper. His research finds that a doubling of carbon dioxide will increase temperatures by only 0.7°C, which is significantly below the lowest estimates of the models used by the IPCC. The climate industry closed ranks it took about two years for Lindzen to find a journal to publish the paper. Critics will of course focus on the point that the paper was published in an Asian, not a Western, journal. Notwithstanding, the scientific facts do not change.

Adolf Hitler sought to discredit the work of Albert Einstein by commissioning a paper called "100 Scientists Against Einstein". Einstein responded: "If I were wrong, one would have been enough".

The media views science as a popularity contest in which those with the greatest numbers win. Consensus is politics, not science. If there is a hypothesis that human emissions create global warming, then in science only one item of evidence is needed to show that this hypothesis is wrong. Dozens of items of evidence have been listed here and in other writings to show that the human-induced global warming hypothesis is wrong. Dominant views in the past were that the Sun rotated around Earth, that burning material was due to a substance called phlogiston, that health was driven by humours, that leeches could cure most diseases, that malaria was caused by bad air, that earthquakes were an act of God, that heavier-than-air machines could never fly, that the continents were fixed and that duodenal ulcers were caused by stress. All these views strongly supported by science at the time have been shown to be wrong.

The skewing of the scientific literature is not healthy for a society that depends upon scientists and scientific literature for trustworthy advice for wise policy decisions. My advice to young climatologists: Do not write a paper questioning the IPCC, the popular paradigm of disastrous climate change or your host climate institute. You will destroy your career prospects.

For the first 300 years after the Royal Society was founded, they adopted a position of aloofness from political debates by refusing to become embroiled in the political controversies of the day. In fact, their journal *Philosophical Transactions of the Royal Society* carried a notice: "It is neither necessary nor desirable for the Society to give an official ruling on scientific issues, for these are settled far more conclusively in the laboratory than in the committee room."

This all changed in the 1960s and the Society gradually became increasingly involved in politics and policy. The last two presidents were especially active and it was only a revolt by Fellows that slightly changed matters. The Society now carries a damaged reputation because of its engagement in political controversies. The Royal Society is not alone. Almost all professional societies are involved in political lobbying, policy formulation and advocacy. Many learned societies, including the Royal Society, are partially funded by government. Those societies to which I belong also have internal ructions when the executives acting against the wishes of members by taking a position on political controversies of the day. The same has happened elsewhere. The American Physical Society played the political game and in 2010 revised its statement on climate policy. In a scathing letter of resignation by one of its most eminent and senior members, Hal Lewis (University of California) commented on the popular paradigm of human-induced climate change:

> It is the greatest and most successful pseudoscientific fraud I have seen in my long life as a physicist.

My criticisms of the climate industry are that much of the data (e.g. temperature and carbon dioxide measurements and corrections) is contentious, that computers have to be tortured to obtain the pre-ordained result, that computer codes are not freely available, that neither the data nor the conclusions are in accord with what we know from the present and the past, that publication is within a closed system that challenges the veracity of the peer-review system, that financial interests are not declared, that the financial rewards for publishing a scary scenario are tempting and that the process of refutation seems to have been overlooked. On a bad hair day, I might even describe the climate industry as unscientific.

Evolution of scientific ideas

What would have happened if, in 1900, we had assembled a few hundred geologists into an International Panel on Continental Drift (IPCD) and asked the question: Do continents drift? The answer amongst geologists would have been a resounding no. If we asked the same question again in 1920 after the great works by Alfred Wegener in 1913 to 1914 were translated from German to English after the 1914-1918 war, the consensus would have been negative. A few geologists might have said maybe, but we need more data and a more plausible mechanism. Even fewer would have stuck their necks out and said that the data supported continental drift. The majority of geologists would have said that continental drift was nonsense and used authority, professional societies and peer-reviewed science to support their position. Wegener and his supporters would have been ridiculed. And they were.

In 1940, the consensus of IPCD would have been no. By then Harold Jeffreys had shown that the mechanism proposed by Wegener was impossible, Germany was again at war and there were strong anti-German feelings. The IPCD would have rejected the theory of continental drift and, by doing so, would have rejected all of the supporting data because the mechanism suggested by Wegener was not plausible. In 1960, the IPCD consensus would still have been no. Maybe a few geologists, geophysicists and geochemists might have said possibly but better data and explanations are required. Again, the majority of geologists would have said that continental drift was nonsense and would have used authority, professional societies and peer-reviewed science to support their position. Arguments such as the "science is settled" would have been used.

If the IPCD were asked the same question in 1980, as a result of the integration of many disciplines of geology, the consensus would be that continents do indeed drift and that the mechanism is known. Even in the late 1970s, opponents of continental drift and plate tectonics ridiculed in public those who embraced the new paradigm. In 2000, if the IPCD were asked the question again, the consensus would be yes. Some might give a cautious yes, until a better theory emerges. Others would argue about the details of plate tectonics and whether this all-embracing theory could explain all observations and, as a consequence, would enjoy the ridicule from their peers.

So too with "climate science", an infant scientific discipline, built from a partial synthesis of other disciplines. Mechanisms of climate science are probably far more complicated than the mechanisms of plate tectonics because they involve complex, non-linear, chaotically interacting systems which in turn involve poorly understood fluid dynamics. It may take many decades before "climate science" can come to a better understanding of climate. In the interim, we have a "consensus", "settled science" but a large number of scientists, especially in non-Western countries, opposing the popular view that human emissions of carbon dioxide drive climate change. Over 100 years, the mythical IPCD completely changed its position. The IPCC has only been alive for 20 years and it may take at least another 100 years before climate is better understood, as happened with continental drift (which evolved into plate tectonics). It was the integration of major disciplines of science that led to the theory of plate tectonics. The evolution of "climate science" needs the synthesis of other scientific areas such as geology. This evolution of understanding will be slowed down because of the financial interests of the IPCC and the behaviour the climate industry, environmentalists and scientific activists.

With the idea of continental drift, there were not huge pots of money involved. Initially there were anti-German views. Only nationalism, egos, reputations and careers were at stake. With "climate science", not only are there huge amounts of research funds sloshing around but the economic implications of a carbon trading scheme, "Carbon Tax" and the transfer of wealth have brought together all sorts of disparate and desperate groups. Science funding opportunities, the rise of the Green Party and financial opportunities in the carbon market place have combined to produce an almost perfect storm of quasi-religious hysteria with all the hypocrisy that one associates with a fundamentalist religion. Fear of the environmental consequences of ever-expanding economies combined with the deterioration in science education encourages people to think of simple solutions to problems (or non-problems). This is amplified by a lazy media that exploits fears to increase sales and can't really be bothered to get to the bottom of a story. There has been a great decrease in science ethics. We now have senior scientists seeking to suppress publication of research that undermines their beliefs. We have senior scientists making statements that the "science is settled" on human-induced climate change regardless of contrary findings of others. We now have previously well-regarded institutions discarding scientists whose work is against popular opinion. The age of open transparent science may be coming to an end.

The history of planet Earth has been ignored with the current popular catastrophist story of human-induced climate change. If large bodies of evidence and history are ignored, then this provides an unbalanced, misleading and deceptive view of global climate. If scientists ignore integrated interdisciplinary empirical evidence, then they have politicised science to gain government favours and they are operating fraudulently. If scientists ignore history, then they do so at their peril.

Why is it that I have little confidence in the data, the methods, conclusions and science communication of scientists in the climate industry and of those who promote the doom-and-gloom story? It is a long list. Climategate; ignoring past climate changes, the rates of past changes and the natural variability of climate; adjustment of primary data to yield the required result; corruption of the peer-review process; exclusion of contrary views from eminent scientists; corruption of both the temperature record and the carbon dioxide record; non-correlation of carbon dioxide with temperature on all time scales; conversion of science and free independent inquiry into political advocacy; dampening or omitting of the validated record of the Roman and Medieval Warmings; creation of the 'hockey stick' *ex nihilo*; demonising of dissent; denial that planet Earth changes by other forces far larger than anything humans can create; failure of computer-generated models; massive vested interests promoting the certainty of a human-induced catastrophe coupled with fraud and hypocrisy and the lack of caution and reserve in making public statements about new scientific findings.

These points show that the gains made in the Renaissance have been lost over the last two decades. Any system that allows questioning of beliefs is an enlightened system yet the climate industry is doing everything to stop the questioning of the basics of climate. Such behaviour as outlined above is not that of scientists, but of paid political thugs. Much of the political and media pressure comes from full-time climate advocates paid to misinform. And who will benefit? The punter, of course, will pay. Banks, traders, brokers and insurance companies are lining up in the shadowy world of totally opaque carbon trading to make astronomical profits. The planet will also pay. Traders and banks have totally different objectives to governments, activists and global warming bureaucrats. No amount of profits made from trading will remove carbon dioxide from the atmosphere. It does not appear that we have learned very much from previous financial crises. The amount of funds transferred to bank profits means less money for genuine environmental programs. Government proposals for a green future are, in reality, proposals for opening the floodgates for the laundering of your tax money.

Science has always self-corrected. However, Western governments have uncritically and dogmatically embraced human-induced global warming as an excuse for increasing taxation, redistributing wealth, eroding freedoms, constraining liberal thinking processes and maintaining power by doing deals with groups allegedly concerned about the environment and constraining liberal thinking processes. Countries like Australia that have now racked up massive debt desperately need a "Carbon Tax" to pay back debt. The rapacious banks are salivating. They have never been presented with such an opportunity given on a plate. And, as we all know, the banks are really concerned about the environment and will not hesitate to churn all their profits back into the environment. These changes take only decades to enact and centuries to reverse following massive economic and human disruption.

Some of the leading lights in the climate industry use a novel way of scientific argument. For example, leading warmist David Karoly (University of Melbourne) dismisses contrary scientific arguments presented by Stewart Franks (University of Newcastle). Rather than addressing any evidence or argument presented by Franks, Karoly makes public statements while claiming that a public response by Franks is unprofessional. Karoly presents himself as a climatologist of high authority and denigrates those with more specialised expertise who do not agree with his pronouncements. He employs an implicit threat of litigation against those such as Franks who question him. A more convincing scientist would handle questioning with data, logic and argument. If radio journalist Alan Jones can chew up and expose David Karoly in a short interview, just imagine his performance under a few days of vigorous cross-examination in Court.

The tragedy of the climate industry, the politicisation of science and the grubby Climategate affair is the weakening of science. We can only feed ourselves because of science. It is science that gives us longer and better lives. When the next inevitable pandemic hits humans, we are not going to solve problems by denigrating those with an opposing view and claiming consensus. Governments may be so embittered by the human-induced global warming scam that there may be no funding for disciplines such as medical science. In a pandemic, we would need science, independent thinking, creativity and for scientists to tread where no one else has been. That is the only way to solve problems but these very problem-solving processes are being undermined and destroyed by the climate industry.

What if I am wrong and a reduction of carbon dioxide emissions is absolutely necessary to save the planet? If Australia stopped all carbon dioxide emissions today, global temperature would decrease by 0.0154°C by 2050. Not only would Australia become bankrupt and could not feed itself, such voluntary acts of international environmental kindness would have absolutely no effect on the global climate. Maybe the Green Party would willingly commit economic suicide, no one else would least of all developing countries like China and India. They want our standard of living and nothing is going to stop them.

Models, predictions and adaptation

Computer climate models throw no new light on climate processes. The science underpinning the hypothesis that humans drive global warming is not in accord with the past. Climate models tell more about the modellers than the climate as they produce pre-ordained conclusions, cannot be run backwards to show what has been validated and have been shown to be wrong by measurements. Climate, atmospheric temperature and ocean temperature models have all been checked with measurements and all models have been shown to be incorrect. Current mathematical models of climate are mere scenario predictions of the assumptions used to drive the model. The fact that these are presented as accurate predictions of real climate rather than extensions of what seems to be unfounded assumptions is dishonest. A model is not scientific evidence.

The key idea of global warming theory is that carbon dioxide emissions should trap heat in the Earth's atmosphere preventing it from escaping into space. This has not been measured, it has been predicted from computer modelling. This heat should be trapped at altitude in tropical areas yet balloons show that no such heat is trapped.

Balloon measurements of heat have been validated. In a paper in *Remote Sensing*, NASA satellite data over the past decade has been used to show that our atmosphere releases much more heat into space than the computer models show. The missing heat is lost to space, not to the oceans and atmosphere as the models predicted. Measurements give a different story. If you have the choice between measurements and a computer model, what would you use?

When NASA satellite data, reported in a peer-reviewed scientific journal, show a "huge discrepancy" between alarmist climate models and real-world facts, climate scientists, the media and our elected officials would be wise to take notice. Whether or not they do so will tell us a great deal about the honesty of the purveyors of global warming alarmism. When five years ago, it was pointed out in the New Zealand press that the planet had not warmed for five years, David Wratt (an IPCC lead author) responded that climate models could not be expected to be accurate in anything less than a 10 year period. Now, after 11 years of cooling, this claim looks even more ridiculous. If computer models of future warming are used, then there must be a huge warming between now and 2020 just to catch up. This would be 0.8°C between 2011 and 2020, twice the fastest rate that has been measured over the last century. Either we can accept that the models are hopelessly wrong or we can continue to believe that there are fairies at the bottom of the garden.

NASA satellite data for 2000 to 2011 shows that the Earth's atmosphere is allowing far more heat to be released into space much faster than computer models predicted and that carbon dioxide in the air traps far less heat than warmists claimed. Scientists on all sides of the global warming debate are in general agreement that very little heat is being directly trapped by human emissions of carbon dioxide. However, the most important issue is whether carbon dioxide emissions will indirectly trap far more heat by causing large increases in atmospheric humidity and cirrus clouds. Computer models assume human carbon dioxide emissions indirectly cause substantial increases in atmospheric humidity and cirrus clouds (both of which are very effective at trapping heat) but this has not been demonstrated.

Humans have adapted to live on ice, in mountains, in the desert, in the tropics and at sea level and can adapt to future changes. History shows that during interglacials, humans create wealth which allows populations to grow whereas glaciation is associated with famine, starvation, disease and depopulation. The cycles of climate change suggest that the next inevitable glaciation will be little different from previous glaciations. Imagine ice sheets covering the same area as in the last glaciation. Most of Europe, Canada and northern USA would be covered by ice. Other areas such as China, Mongolia and Australia would have howling cold winds and shifting desert sands. Alpine areas would be covered with ice and ice sheets would greatly expand. The food supplies for seven billion people would be really under pressure. The best we humans can do is prepare for change (which we have never done in the past) and adapt (which we have done in the past). Because of modern technology, any adaptation to modern climate change would be far easier this time.

The models also claimed that modest warming by carbon dioxide would be amplified by increased water vapour in the atmosphere. This can be tested by checking actual rainfall against the predicted rainfall. Out of eighteen regions tested in USA, eight show that rainfall would either increase or decrease depending upon the model used. One model predicted an 80% decrease in rainfall and another predicted an 80% increase in rainfall. Neither deserts nor swamps appeared. The models had no bearing on reality.

Climate policy is based on models and model projections and not empirical data. Science magazine says that NASA scientists rechecking ice-monitoring data from eastern Antarctica report that the melting of large volumes of ice from the continent has been exaggerated. A study by Phil Watson, a coastal researcher with the New South Wales Office of Environment and Heritage, shows the centuries-long rise in sea levels in eastern Australia has slowed and is now increasing at "a reducing rate". This is contrary to the exaggerated claims of the climate industry.

The Australian climate commissioner Tim Flannery does not seem to be able to get any of his predictions right. Climate Change Minister Combet gave the Climate Commissioner's job description as: "... (to) provide an authoritative, independent source of information on climate change to the Australian community."

Flannery does not provide independent advice to the government, he seems to get every prediction wrong and he refuses to justify his climate alarmism in public. We don't know whether Flannery's predictions are computer-based or are spur-of-the moment statements.

This is in accord with the second part of his job description: "..(to) build the consensus about reducing Australia's carbon pollution."

Maybe Tim Flannery should go back to Papua New Guinea and continue studying his tree kangaroos, his area of expertise. Climate commissioner Tim Flannery has alarmism in his arsenal. His forecasts of doom-and-destruction are regularly lapped up by the non-critical media. But, when given the opportunity to debate those that know something about climate, Flannery heads for the hills. Although he is funded by the public as the climate commissioner, the *Spectator Australia* showed that Flannery refuses to debate or justify his position in public. He was joined in this refusal by many other climate alarmists (Ross Garnaut, Will Steffan, Clive Hamilton). Why? Do they know that their position is untenable? Do they know that they will be exposed for what they are? If there is a consensus and the science for human-induced global warming is so strong, then they should relish opportunities to silence their critics. By not debating critics, one can only conclude the obvious. Some alarmists claim that the science is settled hence debate is not necessary (e.g. Steffan). If that is the case, then there is no more need for climate research and the research institutions, including Steffan's, should be shut as further work would only confirm what they already know.

In an October 2006 opinion piece entitled *Climate's last chance*, Tim Flannery tried to scare readers about sea level change with: "Picture an eight storey building by the beach, then imagine waves lapping its roof."

Tim Flannery lives at sea level in a waterside house. He purchased the adjacent property, also at sea level. Apart from the blatant hypocrisy, he omitted to mention that the most catastrophic sea level predictions are 10 millimeters per year hence it will take thousands of years to reach Flannery's alarmist sea levels. He clearly does not believe his own sea level change predictions and clearly does not believe that there is a problem with global temperature, as he stated on Melbourne Talk Radio (in March 2011) that: "If the world as a whole cut all emissions tomorrow the average temperature of the planet is not going to drop in several hundred years, perhaps as much as a thousand years."

As climate commissioner at the National Climate Change Forum he warned Australian families that their summer beach trips would be a thing of the past: "It's hardly surprising that beaches are going to disappear with climate change."

There has been climate change since the beginning of time, beaches only disappear when there is sea ice. Beaches move with falling and rising sea levels. Flannery should have known this because he lives at sea level on a drowned river system and all along eastern Australia are submerged beaches, raised beaches and inland beaches showing that sea level has been up and down with no catastrophic consequences. In

June 2011, he was at it again: "There are islands in the Torres Straight that are already being evacuated and are feeling the impacts".

This is nonsense. People moved from an island but for other reasons. Furthermore, if sea level is forcing people to leave islands in the Torres Straight, then why are people still living on other islands in Torres Straight and around the coast of Australia.

One of Flannery's 2005 dire predictions was that Sydney's dams could be dry in as little as two years because global warming was drying up the rains. The dams did not dry and in 2011 were at more than 70% capacity. In order to really frighten people in 2007, he claimed just before heavy rains and great floods that many Australian cities will run out of water and that: "Perth will be the 21st century's first ghost metropolis" and that Adelaide, Sydney and Brisbane would: "need desalinated water urgently, possibly in as little as 18 months."

Brisbane dams spent much of 2011 at overcapacity. In 2007, Flannery announced in *New Scientist* that: "Australia is likely to lose its northern rainfall".

It didn't and Australia has had some pretty challenging recent wet seasons. In 2008 he tried to frighten the good people of Adelaide with "the water problem for Adelaide is so severe that it may run out of water by early 2009."

It didn't. And the dams are at more than 75% capacity and a desalination plant is an expensive white elephant. It appears that every time Flannery makes a prediction, the opposite happens yet he seems uneasy about predictions when he states (*Lateline*, November 2009): "So when the computer modelling and the real world data disagree, you've got a very interesting problem".

The obvious way to solve the "interesting problem" is to abandon computer models with their alarmist predictions and stick to real observations. By not abandoning computer models, Flannery allowed ideology to overrule real-world data. *Lateline* foamed at the mouth when earlier in 2009 I showed that the planet was cooling yet carbon dioxide emissions were increasing but they did not question the same science in November 2009 when their darling climate commissioner said the same thing: "Sure, for the last ten years we've gone through a slight cooling trend".

It is the same Tim Flannery who, as climate commissioner getting paid $180,000 per year for three days a week "work", tells us to accept the science regarding global warming. When his science and predictions turn out to be so consistently wrong, when he states that the computer modelling and real-world data don't agree and when he states that there is currently global cooling why should we even listen to him let alone accept the science that he promotes?

The story of planet Earth is a marvellous chronicle written in stone. The only way to understand climate is to read the rocks because the present derives from the past. To view modern weather and climate in isolation from the past is nonsense. And this is exactly what is done by those who claim to be "climate scientists". To make predictions based on computer models can only create wrong answers. There has been testing of enough models to show that not one has predicted what we have experienced or measured. Past climate changes have been very complicated in a chaotic non-linear system with sporadic randomness. These systems are poorly understood and it is only by looking at the past and integrating with what we know about the present that we can hope to understand major natural processes. That understanding is a long way away. For scientists to argue that traces of a trace gas emitted by humans into the atmosphere are the main driving force for climate changes on planet Earth is fraudulent. To argue that every change on a dynamic planet is due to human activity ignores the rich past that the chronicle of planet Earth gives us.

Although the history of planet Earth will always be incomplete, we have enough empirical evidence from history, archaeology, glaciology and geology to show that past climate changes have never been driven by trace additions of carbon dioxide into the atmosphere hence there is no reason to conclude that present human emissions of carbon dioxide will be any different.

Anti-science

Most scientists concentrate on their science. Some are involved in teaching. A few promote science to the broader community. Only a handful of active scientists actually deal with the ingress of anti-science into the schools and community. If litigation, broadcasting, writings and public lectures are any guide[5], then I am one of those few scientists who tackles anti-science. I am more than qualified than any scientist in Australia to make comments on, for instance, creationism.

Creationists, as a matter of faith, believe that planet Earth is a mere 6,000 years old, that all sedimentary rocks and fossils were formed in a great flood some 4,000 years ago and that evolution does not exist. Creationists invented "creation science" and they called their evangelical ministers "creation scientists". Some of these "creation scientists" have basic scientific qualifications. The "science" of creationists is presented with scriptural authority and intertwined with morality. Creationists have tried to have their narrow fundamentalist view of the world taught in school science courses and have exerted considerable political pressure to be given a fair go and be allowed to teach an alternative to evolution to every school child in a pluralist society.

There is an uncanny similarity between creationists and the climate industry. Just as creationists invented "creation science", the climate industry invented "climate science". Both areas use a very narrow view of science to come to an all-embracing world-view. Both groups have their prophets (e.g. Duane Gish, Al Gore), a few prominent leaders (e.g. Carl Wieland, Tim Flannery) and an army of the faithful worker bees (e.g. Phil Jones, Michael Mann; Andrew Snelling, Ken Ham). Both groups have a holy book (the *Bible*; the latest IPCC reports) but very few of the faithful have actually read and understood their holy books. The faithful just believe. Both groups publish their own work

5 Ian Plimer, 1994: *Telling lies for God: Reason vs creationism* (Random House).

in their own journals over which they exert strong editorial control. No work that challenges the pre-conceived dogma ever sees the light of day in their journals. In their various small circles, the officers of both movements profit in terms of money, power and prestige. With both groups, there is a strong underpinning the principles of Christianity. With the creationists, salvation can be purchased by donations, purchases of literature and DVDs. Both creationists and global warmers have a catastrophist view of the world. They both incorporate aspects of catastrophic scenarios which various unbalanced folk have touted off and on for thousands of years. A negative view of the future predominates. However, all is not lost. Salvation can be purchased (e.g. carbon tax, emissions trading) and, once indulgences have been purchased, then the planet or your soul will be saved.

Both groups are long on cant and short on fact, intertwine data with interpretation and are totally unwilling to expose their data to scrutiny. The parallels become even closer when one looks at the use of data. Climate catastrophists and creationists use a very narrow body of data, wallow in trivial obscure science, argue that any natural variation proves their point and cannot integrate data from all disciplines of science to get an overview of the holistic workings of planet Earth. Creationists are quite happy to read only the creationist literature, controlled by their masters. "Climate scientists" are quite happy to only read the climate literature that is tightly controlled by the climate club set up to maintain a consensus. Both groups do not display a broad knowledge, have no well-read polymaths and can only stick to their group-think message. At the rare times when there is public debate, both groups make expressions of faith and morality, paint doomsday scenarios and they cannot present the simple science that underpins their dogma.

The arguments of both groups fail when confronted with geology, evolution and extinction of life and sea level changes. Time unstitches both the climate industry and creationists. Creationists use fraud to try to show that radioactive dating of rocks demonstrates that planet Earth is only a few thousands of years old rather than 4,500 million years old. Climate catastrophists also totally ignore time. Geology is excluded from climate debates because it shows that past changes in climate have been greater and quicker than anything measured today and are not related to atmospheric carbon dioxide. Climate catastrophists only focus on the last 50 years to predict what will happen hundreds of years into the future. Climate catastrophists, like creationists, normally don't collect the primary data themselves. They massage the data of others and, in the case of climate catastrophists, use intricate computer models to try to frighten their flock. Inconvenient measurements are conveniently ignored. Creationists ignore geological data whereas various climate centres (e.g. Hadley) ignore large bodies of data (e.g. Russian temperature data) that don't back their preconceived story. Others (e.g. CRU) keep their data in such a chaotic mess that nothing can be checked.

Messages from God abound. Any variation in a natural process over the last 100 years is, of course, due to the action of humans or God, depending upon whether the speaker is a climate catastrophist or a creationist. Even when shown to be scientifically wrong, both groups still cling to their treasured theories. Creationists and the climate industry have lectures, public meetings and conferences that are closed shops. No one who could challenge their views in front of others is invited or can attend. Both groups need something to fear and become catatonic if someone tries to take away their fears, no matter how irrational.

History is ignored by both. With the climate industry, history shows that previous warmings have been longer and more intense yet non-catastrophic than the very slight late 20th century warming. The past warmings could not possibly have been driven by industry belching out carbon dioxide because, unless all of history is wrong, there was no heavy industry in past periods of warming. History is also pretty cruel to creationists because it appears that there were empires, towns and villages on planet Earth well before their Earth was created, that many communities were not wiped out by a great global flood and that all sedimentary rocks and fossils could not possibly have formed in this flood. Both climate catastrophists and creationists just do not accept history. Creationists try to change dates such to make history agree with their view of the world; similarly the Medieval Warming was simply removed from the record by Michael Mann and the IPCC.

For both groups, carbon 14 causes huge problems. Creationists are a little more sophisticated than the climate industry and use arguments about the accuracy of carbon 14 dating to validate their view of history. They have trawled through the scientific literature and found the occasional published radioactive date that is wrong and they have attacked the methodology of radioactive age dating. The climate industry simply ignores carbon 14 evidence that shows changes in solar activity and cosmic rays.

Both climate catastrophist and creationist dogma can be easily destroyed with a traverse through a thick section of layered sedimentary rocks which show that planet Earth is old, that sea level constantly changes, that climate rapidly changes from icehouse to greenhouse conditions, that climate is never static, that planet Earth for most of time has been warmer than now and that life survives these great changes. This is not new. It was done in the 18th century and has been validated thousands of times. But, why should either group use empirical evidence when belief is a matter of faith? This, of course, is a waste of time. No matter how many times the climate catastrophist and creationist dogma is shown to be wrong, their dogma, faith, financial interests and career prospects just do not allow them to abandon a treasured and lucrative theory.

I somewhat modestly claim that I am an expert on Noah's Ark. It is a dubious honour but someone has to carry this burden. This expertise is supported by various writings, broadcasts, television programs at Mt Ararat and litigation regarding Noah's Ark. The

Noah's Ark message is that God destroyed most life on Earth with a massive worldwide flood because we were not behaving ourselves. The good guys and a few species carried on the Ark survived because they were sensible enough to fear and listen to God. Sinners and most life died. We see the same moralistic thinking among climate catastrophists. We humans are evil, there are messages sent in the atmosphere, life and oceans by Gaia and we need to change our ways otherwise we are finished. Again, fear and guilt are the weapons used to try to force us to change our ways. If we don't pay up and change our life, then we will fry-and-die, sea level will rise and there will be massive extinctions of life. There might just be space for us on planet Ark but a berth will cost. This Noachian religious remnant underpins much of the environmental movement and has been embraced by scientists profiting from the catastrophist climate cause. The Ark is the environmentalists' religious symbol for biodiversity.

Both groups redefine science, have a very selective use of data, are unable to revise their cherished theories and become aggressive when their opinions are challenged with contrary evidence. There is a chronic aura of insecurity in both groups who feel that the Earth was once perfect and static in either the Garden of Eden or pre-industrial times. Both groups believe humans have destroyed the world by sin. This shows that both groups are essentially anti-human.

Both movements have all the trappings of humourless Western fundamentalist cults who intertwine their theology with a world-view replete with morality and both argue that science underpins their worldview. Creationists have little interest in theology, just as the climate industry has little interest in the environment. It is a game of power. The power games involve mutual support groups, shutting out of dissidents and threats to apostates of total destruction, either as eternity in hell or the destruction of a career. The climate industry, like creationism, ignores data, creates data *ex nihilo,* "adjusts" data, selectively uses data and cooks the books, albeit in different ways. This unifying thread of deceit ties both groups together. Why is there the necessity for both groups to adjust raw data?

When I was challenging the creationists in public, I was the darling of the left. They incorrectly believed that I was attacking Christianity. I was not. I was attacking fundamentalists who exploited Christianity to claim that science underpins their scientifically incorrect worldview. All mainstream Christian religions supported me. I attack the climate industry because they exploit people's yearning for spirituality by claiming that science underpins their scientifically incorrect worldview. Now the loopy left attacks me because their non-scientific view of the world is anti-industrial. I attack both creationism and the climate industry because they abuse science. Some scientists (e.g. Michael Archer) criticise creationism using the coherence criterion yet are unable to use the same logic for questioning human-induced global warming. Maybe they are too entrenched in the climate industry to see the wood for the trees.

The climate catastrophist movement has been embraced by much of society because they are either not scientifically educated or they need to believe in something. It is the new fundamentalist religion of the West that has replaced Christianity and has many elements of the replaced religion (sin, indulgences, salvation, worship of a higher authority). However, the new environmentalist religion is atheistic, anti-human, vacuous and secular with no music, history, literature, coherent philosophy or deep thinking. Just scratch the surface of a Green Party politician and this becomes evident very quickly. Just scratch the surface of a creationist or a climate catastrophist, and you will find a very angry person.

I see very little difference between creationism and climate catastrophism. We could satisfy our spiritual needs with music, history, literature, thinking and a coherent spiritual philosophy rather than a vacuous environmental quasi-religious fad to save the world.

3

CARBON DIOXIDE

Planetary degassing and carbon dioxide

Since the formation of planet Earth, there has been degassing to the atmosphere of water vapour, carbon dioxide, methane, sulphur-bearing gases, nitrogen and many other gases by the processes of plutonism, volcanism and metamorphism. Degassing occurs on other planets, as does climate change. Degassing of carbon dioxide occurs before, during and after volcanic eruptions from gas vents, hot springs and craters. Most degassing occurs before eruptions. Some 1,500 terrestrial volcanoes are known and only two-dozen or so are accurately monitored. Measurement of carbon dioxide vented from volcanoes is not accurate as only atmospheric carbon dioxide is measured and not carbon dioxide dissolved in steam.

In a laboratory, when rocks are cooked up with gas, liquid rock contains far more dissolved gas than natural liquid rocks at the surface and the gas lowers the melting point of the rock. Liquid rocks at the surface are at this lower melting temperature showing they have lost gas. Measurement of volcanic carbon dioxide during eruptions occurs well after the major degassing and deposition of carbon dioxide-bearing minerals. Measurement of carbon dioxide in an eruption does not measure the total amount of carbon dioxide released from a volcano.

Submarine degassing occurs from at least 3.47 million off-axis submarine basaltic volcanoes and from almost continuous gas venting along the 64,000 km strike length of the mid-ocean ridges. At mid-ocean ridges, the oceanic crust of the Earth is pulled apart and gases from deep in the mantle leak to the surface. Each year some 10,000 cubic kilometres of seawater circulates through new mid-ocean ridge basalt as a coolant. This heats the ocean. Experimental studies show that carbon dioxide is highly soluble in sea floor volcanic rocks (i.e. basalt) as compared to common terrestrial volcanic rocks (i.e. andesite, rhyolite). Quickly frozen basalt collected from the sea floor, although containing carbon dioxide, tells us that most of the carbon dioxide dissolved in basalt is vented before eruption.

In 1999, a slow spreading mid-ocean ridge (Gakkal Ridge, in the Arctic Ocean) experienced an explosive submarine basaltic eruption. For basalt to explode at such a great water depth, at least 13.5% of the molten rock must have been carbon dioxide. The new volcanic rocks needed to be cooled by circulating seawater and the Arctic Ocean warmed for a short time. This warming was coincidental with a lunar tidal node that pushed warmer surface North Atlantic Ocean water into the Arctic.

There are no deep submarine measuring stations. Hence the emissions of heat and carbon dioxide from submarine basaltic volcanism can only be deduced from other areas of geology. In some places, liquid carbon dioxide has been found on the ocean floor and gas vents of carbon dioxide are very common. Furthermore, submarine carbon dioxide released from gas vents, hot springs and submarine basalt eruptions dissolves in cool high-pressure bottom waters for degassing to the atmosphere thousands of years later during upwelling and mixing. Some 97% of annual emissions of carbon dioxide are natural and much of the natural emissions derive from the oceans degassing volcanic carbon dioxide. I stress that this oceanic volcanic degassing is a continuous process related to but not necessarily controlled by the sporadic volcanic eruptions.

Degassing also occurs from rising bodies of molten rock that freeze kilometres below the surface (i.e. plutons). During ascent, they undergo constant degassing of steam, carbon dioxide and methane as do wet sediments and limey rocks. In places, these gases are used to drive geothermal power stations. Plutonic and volcanic rocks mainly occur where areas of the Earth's crust are pushed together, such as the collision of Africa with Europe. In this setting, mountains also form (e.g. European Alps) and, as rocks are compressed, gases such as steam, carbon dioxide and methane are released. These commonly form the spa waters typical of alpine areas. With both plutonic and volcanic activity, large amounts of carbonate minerals are deposited in cracks and fissures. These are of interest to geologists, because such fissures often contain gold and silver and hence have been investigated in detail. They are also the fingerprints for carbon dioxide degassing. In discussions in the scientific literature on emissions of carbon dioxide, the number and magnitude of carbon dioxide emitting submarine volcanoes is ignored as are other sources of carbon dioxide.

The carbon isotope chemistry of carbon dioxide dissolved in the oceans from past volcanic activity is the same as that released from the burning of fossil fuels. The main method for determining the amount of fossil fuel burned uses carbon isotope chemistry and hence the method may be measuring a combination of human and natural emissions of carbon dioxide. The method is used because it is not possible to determine how much carbon dioxide is generated globally from the burning of fossil fuels because much carbon dioxide is emitted by villagers burning poor quality coals in Third World countries.

Carbon dioxide is not pollution. Real pollution derived from rapid uncontrolled post-war industrial expansion spawned two environmental movements. The first consisted of physical scientists and engineers who directly addressed pollution by developing facilities and legislative controls. This has virtually eliminated industrial pollution in the Western world. The second comprised activists with little or no physical science background. They did nothing but protest against industry, didn't create a single pragmatic solution to an environmental problem yet bathe in symbolic victories.

An innocent trace gas

Carbon dioxide is a greenhouse gas. The main greenhouse gas is water vapour. Both help to keep the planet habitable otherwise all water would either freeze or vaporise. The first 100 parts per million carbon dioxide in the atmosphere is vital. It helps to keep the atmosphere warm. After that, any more carbon dioxide in the atmosphere has a diminishing effect. The atmosphere currently has less than 400 parts per million carbon dioxide. Human emissions of carbon dioxide need to be placed in perspective. If annual emissions of carbon dioxide to the atmosphere comprise 33 molecules, only one is from human emissions and the rest are from natural processes. This one molecule of human-derived carbon dioxide is within 88,000 other molecules in air. If human emissions of carbon dioxide drive climate change, then it has to be demonstrated that this one molecule in 88,000 drives climate change and that the 32 molecules of carbon dioxide derived from natural processes do not.

There are many analogies floating round in cyber space. One of the best looks at a length of air at sea level 1 kilometre long. In this 1,000-metre length, 770 metres would be nitrogen, 210 metres would be oxygen, water vapour would be 10 metres, argon 1 metre, and the rest of the gases 1 metre of which carbon dioxide would be 0.38 metres. The human emissions of carbon dioxide would be 1 millimetre and the Australian component of carbon dioxide for the 1,000 metre length of air would be 0.015 millimetres, the thickness of a human hair. And it is this 0.015 millimetre – this thickness of a human hair in a kilometre - which Australian politicians and the green industry call pollution. Why are we wasting so much money and effort about such a trivial issue when there are far more pressing issues for the nation to address?

The total amount of carbon as carbon dioxide in the atmosphere is 720 gigatonne (Gt). This seems like a very large amount. It is. A very conservative estimate of carbon in soil, vegetation and biomass is 2,000 to 3,000 Gt whereas the oceans contain about 38,000 Gt. The natural carbon cycle removes carbon dioxide from the air into the oceans, as it has always done. Even the IPCC shows that more carbon dioxide enters the oceans than leaves. As with all matters of science, unless you happen to be talking to a "climate scientist", there are large uncertainties. An unknown but huge amount of volcanic emissions of carbon dioxide may appear thousands of years later as ocean emissions and we really do not know the volume of greenhouse gases emitted by micro-organisms in the top 5 kilometres of the Earth's crust. We do know that these organisms are the greatest biomass on Earth. All this needs again to be placed in perspective. Some 65,000,000 Gt of carbon as carbon dioxide is held in limestone (calcium carbonate) and an even greater amount resides in sediments, sedimentary rocks and the Earth's mantle.

Models of carbon dioxide emissions do not incorporate all variables and uncertainties. For one thing, climate models neglect cloudiness which may be a vital factor for warming and cooling. No wonder models have been shown to be wrong with even short-term predictions because climate is like a game of chess. There are a few pieces yet there are an infinite variety of situations that can derive from their positioning.

It has yet to be shown that human emissions of carbon dioxide drive climate change. In fact, there is only evidence to the contrary. If there really is an effect on global temperature from human emissions of carbon dioxide into the atmosphere, then it is lost within the variability and noise of natural climate change.

We are children of the Sun. The Sun drives the water cycle and the carbon cycle depends upon the water cycle. Climate is modulated by water vapour, the main greenhouse gas. Solar energy, carbon dioxide and water are used by plants in photosynthesis. Agricultural experiments show that drought- and heat-stressed plants thrive and use less water when there is more carbon dioxide in the air.

It is clear that the sources and sinks of carbon dioxide are poorly understood, as are feedbacks. If there were a straight-line relationship between carbon dioxide and temperature, there should be a 1°C increase for every additional 100 parts per million carbon dioxide in the atmosphere. There is not because the relationship is a curve, not a straight line. Therefore, if the current carbon dioxide content doubled, atmospheric temperature would only increase by 0.2°C. If it quadrupled, temperature would increase by a further 0.1°C.

If we burned all the known fossil fuel on Earth, the atmospheric carbon dioxide content would not double. This assumes that no carbon dioxide is dissolved in oceans or used by life. There is a perception that the atmosphere has a huge content of carbon dioxide. Not so. In the geological past, the atmosphere contained more than 30% carbon dioxide compared to less than 0.04% today. Yet there was no catastrophic global warming then. In fact, each of the six major ice ages was initiated at a time when the atmospheric carbon dioxide content was higher than at present. As atmospheric carbon dioxide increases, more and more dissolves in the ocean waters, a feature known by brewers for a thousand years. Ice core measurements show that with past interglacials, temperature rose some 800 to 2,000 years before the atmospheric carbon dioxide content rose. All this is telling us that, although carbon dioxide is a trace gas in the atmosphere and a minor greenhouse gas, it does not drive modern global warming.

In more modern times, planet Earth enjoyed the Roman and Medieval Warmings when there were 600 and 400 years respectively of times far warmer than now. This is well-documented history. These two warmings were separated by the cold Dark Ages when glaciers advanced, crops failed, famine was rife, the weakened population succumbed to the plague and there was massive depopulation. There was even cannibalism as European populations starved. During the Roman and Medieval Warmings, glaciers retreated but sea level did not rise drastically, there was no sudden ejection of carbon dioxide into the atmosphere from industrialisation and it was so warm that the Romans were able to grow crops at latitudes where such cultivation is now not possible. Hannibal was able to take elephants over the ice-free Alps. In the Medieval Warming, the Vikings grew barley, wheat, sheep and cattle on Greenland in areas where farming is now impossible.

The Little Ice Age started in 1300 AD. The coldest periods were those of no sunspot activity (i.e. the solar magnetic field was subdued and allowed more cosmic radiation to hit Earth). Temperatures fluctuated wildly, there were short warm periods interspersed with long cold periods, glaciers advanced and retreated slightly in warmer times, crops failed, people starved, the plague struck Europe in 1347 AD and there was massive depopulation. The coldest period, the Maunder Minimum, was 330 years ago. Since then, planet Earth has been warming. If the planet has been warming for over 330 years, which part of this warming was natural and which part is human-induced? Since thermometer measurements were recorded, temperature decreases (1880-1910, 1940-1977, 1998-present) and increases (1860-1880, 1910-1940, 1977-1998) show that there is no correlation of temperature with atmospheric carbon dioxide. With no correlation between global warming and atmospheric carbon dioxide on geological, ice core and historical time scales, there can be no causation. The pattern of Pacific Decadal Oscillations during the past century matches warmings and coolings in a long-term warming trend.

Ice on Earth is rare. Water vapour has been the main greenhouse gas for all of recorded history. For more than 80% of its lifetime, planet Earth has been warmer and wetter than at present and, since about 2,500 million years ago, the atmospheric carbon dioxide content has decreased from ~30% to the present 0.039%. The decrease in carbon dioxide results from the long-term sequestration into carbonate rocks by life. Sediments and altered rocks also sequester huge amounts of carbon dioxide. In former times of high atmospheric carbon dioxide, oceans were not acid, there was no runaway greenhouse and the rate of change of temperature, sea level and ice waxing and waning was no different from the present. The oceans have been alkaline for all of Earth's lifetime because the chemistry of seawater, ocean floor sediments and new volcanic rocks on the sea floor buffer seawater to stop it from becoming acid, even during times of carbon dioxide concentrations that were thousands of times the present value. When we run out of rocks on the sea floor, the oceans will become acid. Don't wait up.

Because the atmosphere has greater than 100 parts per million carbon dioxide, a doubling or quadrupling of human emissions of carbon dioxide will have very little effect on temperature unless atmospheric carbon dioxide residence times change to two orders of magnitude higher than past times. Leave carbon dioxide alone. It is only following what water vapour does which, in turn, is following the energy output of the Sun. Those westerners who would want to try to reduce plant food and hence contribute to starvation in the Third World need to consider their moral position.

Currently the main global carbon dioxide measuring station is on top of a mountain in Hawaii. There are other stations (e.g. South Pole, Cape Grim, La Jolla, Point Barrow, Shetland Islands). There were 19 measuring stations established in Europe to measure carbon dioxide for five years (1955-1960) with an accurate state-of-the-art wet chemical method. Despite the expectations that there would be an increase in carbon dioxide due to increasing industrialisation, there was no change in the atmospheric carbon dioxide content. During this time, atmospheric carbon dioxide measurements were taken in Hawaii and Antarctica using the instrumental infrared analytical method. The Hawaiian station was established to measure: "the rise in atmospheric carbon dioxide resulting from fossil fuel combustion."

The Hawaiian station was established to prove a pre-determined conclusion and not to test a hypothesis or acquire a body of data requiring explanation. What the data shows is that the seasonal variations are far greater than the annual increase in atmospheric carbon dioxide and that there is a lag between an increase in temperature and the seasonal increase in atmospheric carbon dioxide.

About 25 times as much carbon dioxide is emitted by humans in the Northern Hemisphere than in the Southern Hemisphere. These emissions occur essentially between the latitudes of 30° and 50° North. Does the Northern Hemisphere carbon dioxide mix with the Southern Hemisphere carbon dioxide to give a global atmospheric carbon dioxide content?

Clues come from the nuclear industry. The only source of the gas krypton 85 in the atmosphere is from nuclear reprocessing plants, reactors and bombs. Most reprocessing plants, reactors and bomb testing were in the Northern Hemisphere. A very distinct change in krypton 85 is evident from the Northern Hemisphere to the Southern Hemisphere showing that gases from each hemisphere do not mix very easily. If krypton 85 does not mix quickly from one hemisphere to another, why would carbon dioxide from the industrialised Northern Hemisphere mix with the Southern Hemisphere atmosphere? The evidence for very slow mixing is reinforced by measurement of carbon 14 created by nuclear weapons in the 1950s and 1960s. This shows that it took many years for exchange of carbon 14 between the hemispheres.

However, measurements of increasing carbon dioxide from Hawaii (Northern Hemisphere) and the South Pole shows that there is no time lag difference between hemispheres in the increase of carbon dioxide in the atmosphere. This strongly suggests that ocean degassing in both hemispheres is dominating carbon dioxide increases. Furthermore, we estimate the amount of carbon dioxide emitted by human activities but we cannot accurately measure the amount of carbon dioxide emitted by natural processes. In fact, the range of uncertainty is larger than the actual amount of carbon dioxide emitted by humans.

Curiously, satellite measurement of carbon dioxide globally does not find sources concentrated where we would expect them in industry and population centres of Western Europe, USA and China. Instead, the carbon dioxide sources appear to be in places like the Amazon Basin, south-east Asia and tropical Africa, where human emissions are minimal. Furthermore, the Mauna Loa (Hawaii) measurements show annual change or increases in atmospheric carbon dioxide by up to 3 parts per million. Some years there were no changes. If emissions were from human activity, every year should show an increase correlating with human emissions. It does not. Murry Salby (Macquarie University) who has used carbon isotopes to investigate this enigma and concludes that man-made emissions only have a small effect on global carbon dioxide levels and states that: "anyone who thinks the science is settled in this subject is in fantasia".

He was once an IPCC reviewer and comments that if this was known in 2007 then: "the IPCC could not have drawn the conclusions it did."

Another innocent trace gas

We hear that cattle burp out methane. In fact, almost all animals emit methane as part of bacteria digesting food. Termites are the major emitters. What we don't hear is information on natural emissions of methane. Methane is continually produced by microbes that decompose plants. Soils, sediments and sedimentary rocks emit methane and a huge amount of methane is emitted by bacteria living in rocks beneath our feet. Coal, oil, hot springs, artesian waters and groundwater emit methane, as do volcanoes. Reaction between coal, bacteria and water produces methane and spaces in old coal mines are filled with methane. Organic matter in sediments at the bottom of the sea decomposes to methane, which seeps from depth into porous rock methane reservoirs that are mainly along continental margins. With slight changes in temperature and pressure at continental margins, this methane can suddenly be released to the atmosphere. This happens frequently. Recent records show that the atmospheric methane content peaks during El Niño times.

One thing we can agree on is that methane is a very powerful greenhouse gas. Right. However, it oxidises in air very quickly to carbon dioxide and water vapour. Right. Methane emissions took off like a rocket in the 20th century. Right. Cattle and other domestic animals emit large amounts of methane. Right. There are more cattle now than 100 years ago. Right. And covering vegetation with water in large dams results in plant decay and the emission of methane. Right. And we humans must stop eating meat and building dams otherwise we will fry-and-die because we are adding methane to the atmosphere. Wrong.

Even the IPCC states in its 2007 Fourth Assessment Report:

> The total global CH_4 [methane] source is relatively well known but the strength of each source component and their trends are not.

And:

> The reasons for the decrease in the atmospheric CH_4 growth rate and the implications for future changes in the atmosphere burden are not understood but are clearly related to changes in the imbalances between CH_4 sources and sinks.

The IPCC use many words to state that they just don't have a clue where methane comes from and where it goes. Why did atmospheric methane start to drop like a bomb in the late 20th century? What happened? Does this mean that we can't use methane any longer to scare people. What went wrong? The answer is simple: nothing. Humans did something right. In August 2011, *Nature* published two contrasting papers on methane. Both papers were peer-reviewed and had different conclusions about methane. Peer-review does not produce a consensus. One paper suggested that there were reduced methane emissions from fossil fuel production and the other paper suggested that there were reduced natural emissions, especially from rice paddies.

The CSIRO has tried to explain the methane enigma:

> It has been a long-held view that the rate of growth of methane

peaked in the 1960s and has since declined. ... The rise coincided
with the massive release of methane in the 1960s from the great
acceleration in the search for oil at a time when methane was seen
as a waste product and flared into the atmosphere.

Over the last two decades, fuel companies recognised its value
and developed the technology to capture it and sell it as natural gas.

This is nonsense. Exploration wells, whether dry or successful are
capped until development. Production wells burn the methane-rich
gas to carbon dioxide and water vapour. The only country that did
not always follow this procedure was the old Soviet Union where gas
might or might not have been flared into the atmosphere.

Contrary to the CSIRO opinion, natural methane was first used
in USA in the 1930s as a substitute for reticulated gas (town gas) and
has been used domestically since then. In the 1960s the Soviet Union
was the second largest consumer of natural gas. Oil companies were
developing major gas fields in the North Sea in the 1970s to supply
European countries. The technology had all been developed in the
1950s and 1960s and not in the last two decades.

In the late 19th and early 20th century, cast iron pipes were joined
every few metres and often leaked methane-bearing town gas which was
derived from converting coal to coke in gasworks. We no longer produce
reticulated coal gas and now use reticulated natural gas. Although we are
using far more natural gas than 50 years ago, we have stopped natural gas
pipelines leaking (especially in Russia). In June 1982, the Soviet Union
enjoyed a massive explosion equivalent to 3,000 tonnes of TNT when
leaked gas from the Trans-Siberian pipeline exploded and burned. This
was the largest non-nuclear explosion on Earth and the fire could be
seen from space. In the communist Soviet Union, the bottom line was
not too important. Since 1995, Russia has been selling gas to Europe,
the consumption has risen ten-fold and profits leaking from pipelines
to the air have been stopped. It was capitalism that reduced human
emissions of methane to the atmosphere, not environmental activism,
taxation or nature.

The CSIRO either has a surreal sense of humour or does not understand its own data. The CSIRO has predicted a 60% increase in atmospheric methane and the Department of Climate Change and Energy Efficiency wants to tax methane. Australia would be the only country with a methane tax. If atmospheric methane decreases, will there be a tax rebate and do bureaucrats in the Department of Climate Change and Energy Efficiency get sacked because they uncritically accepted activist advice from the CSIRO and got policy wrong? Despite the IPCC stating that it hasn't a clue about methane, their scenarios include atmospheric methane increasing. It appears that contrary data can be ignored when the aim is to scare and tax people.

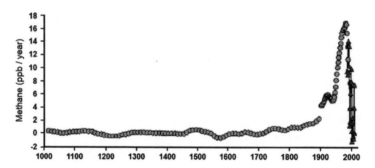

Figure 4: *The rise and fall of atmospheric methane compiled from ice cores (circles, Law Dome, Antarctica) and atmospheric measurement (triangles, Cape Grim, Australia).*

Water and ice

Ice sheets grow and shrink. At times, they disappear. At other times, ice starts to cover polar areas and high mountains. That's what ice has done over the history of our planet. The Greenland and Antarctic basins are more than a kilometre deep and deeper in the centres than around the edges so that ice is squeezed uphill like toothpaste out of a tube by the weight of overlying ice. The alarmist media stresses that changing sea ice and continental glaciers indicate rapid global warming. Is this really so?

Since the last interglacial started some 10,500 years ago, summer sea ice in the Arctic has been far from constant. Sea ice comes and goes without leaving a clear record. For this reason, our knowledge about its variations and extent was limited before we had satellite surveillance or observations from aeroplanes and ships. A huge amount of the Earth's surface water moves alternately between the ice sheets and the oceans. Svend Funder, commenting on his recent *Science* paper stated:

> Our studies show that there have been large fluctuations in the amount of summer sea ice during the last 10,000 years. During the so-called Holocene Climate Optimum, from approximately 8000 to 5000 years ago, when the temperatures were somewhat warmer than today, there was significantly less sea ice in the Arctic Ocean, probably less than 50% of the summer 2007 coverage, which is absolutely lowest on record. Our studies also show that when the ice disappears in one area, it may accumulate in another. We have discovered this by comparing our results with observations from northern Canada. While the amount of sea ice decreased in northern Greenland, it increased in Canada. This is probably due to changes in the prevailing wind systems. This factor has not been sufficiently taken into account when forecasting the imminent disappearance of sea ice in the Arctic Ocean.

In order to reach their unsurprising conclusions, Funder and the rest of the team organised several expeditions to Peary Land in northern Greenland. Funder said:

> Our key to the mystery of the extent of sea ice during earlier epochs lies in the driftwood we found along the coast. One might think that it had floated across sea, but such a journey takes several years, and driftwood would not be able to stay afloat for that long. The driftwood is from the outset embedded in sea ice, and reaches the north Greenland coast along with it. The amount of driftwood

therefore indicates how much multiyear sea ice there was in the ocean back then. And this is precisely the type of ice that is in danger of disappearing today.

What is interesting about this study is that the new understanding came from getting away from computer modelling and doing field work in pretty inhospitable areas. Back in the laboratory and again away from computer models, the wood type was determined and dated using carbon 14. This wood came from near the great rivers of present-day North America and Siberia. This shows that wind and current directions have changed. The field study of coastal beach ridges shows that at times there were waves breaking unhindered by ice over at least 500 kilometres of coastline. At other times due to sea ice cover, there were no beaches. This is the present situation.

Even if there is a great reduction in sea ice, all is not lost. Funder stated:

> Our studies show that there are great natural variations in the amount of Arctic sea ice. The bad news is that there is a clear connection between temperature and the amount of sea ice. And there is no doubt that continued global warming will lead to a reduction in the amount of summer sea ice in the Arctic Ocean. The good news is that even with a reduction to less than 50% of the current amount of sea ice the ice will not reach a point of no return: a level where the ice no longer can regenerate itself even if the climate was to return to cooler temperatures. Finally, our studies show that the changes to a large degree are caused by the effect that temperature has on the prevailing wind systems. This has not been sufficiently taken into account when forecasting the imminent disappearance of the ice, as often portrayed in the media.

Those playing with computer climate models need to get outside, collect new data and take into account far more factors than they feed into computer models.

Studies of the behaviour of tropical glaciers over the last 11,000 years show irregular shrinkage with slower rates in the Little Ice Age and faster rates in the 20th century. Glaciers such as the Bolivian Telata glacier reflects long-term warming during the current 10,500-year long interglacial and that glacial retreat was in progress thousands of years before industrialisation.

Scientists urged on by the media state that ice calving off glaciers indicates global warming. Ice always falls off the front of a glacier. If ice did not melt, then the planet would now be covered in ice. Ice drops off the toe of both advancing and retreating glaciers and the melting snout of a glacier is at a point determined by the balance between the forward movement of the ice by gravity and the rate at which it melts. Ice falling off the front of a glacier means absolutely nothing when the air temperature is less than zero. Ice sheets grow and contract. At times, ice sheets disappear. The story of glacial retreat of is far more complex than a simple television image. Many glaciers that are now in retreat did not exist until the Little Ice Age (which climaxed in the middle to late 17th century). During the Medieval Warming (which peaked at about 1000 AD), alpine glaciers in the Northern Hemisphere were either smaller or did not exist. Over much of the Canadian Cordillera, there may have been no glaciers at all during the Holocene Maximum (8,000 to 6,500 years ago), a period when temperatures were considerably higher than now. Records from New Zealand and Norway show that glacier retreat commenced in the 18th and 19th Centuries. Most of the modern ice retreat is due to post-Little Ice Age warming, changes in humidity and a decrease in ice flow rates.

The idea that a glacier slides downhill on a base lubricated by melt water was a good idea when first presented by de Saussure in 1779. We now know a lot more yet this treasured idea remains. Ice moves by creep, a process of constant recrystallisation of ice crystals. Ice at the snout of a glacier has crystals 1000 times larger than those in snow as a result of growth during recrystallisation. Ice sheets in Antarctica and Greenland first flow uphill before flowing down glaciers. The upward flow of ice cannot be due to human-induced global warming producing melting. There are some places in the world today where glaciers are expanding. Ice sheets and glaciers grow and retreat for a great diversity of reasons. This is what ice does. For scientists to argue that ice retreat is due to human activity is simplifying a very complex process. Furthermore, it is too cold in Antarctica and Greenland for ice to melt.

Since the discovery of the Hubbard Glacier (in Alaska) in 1895, it has been advancing 25 metres a year during periods of cooling and warming. The ice front is 10 kilometres long and 27 metres high. What does the ice do at the snout of the glacier? It falls off because it is getting pushed from behind. This has nothing to do with temperature, it shows that ice behaves as both a plastic and brittle material and that ice sheets are always changing.

As with all areas of science, there are regular surprises. It was always thought that ice formed from frozen snow. The science was settled and there was a consensus. Recent work in the East Antarctic shows that the deepest part of the ice sheet contains ice that did not originate as snow. It was melt water that seeped to the base of the ice sheet and then froze. The amount of ice formed by this method is probably greater in volume than all glaciers on Earth outside Antarctica and Greenland. The computer models predicted that this melt water escaped to the oceans and contributed to sea level rise. Wrong. The volume of water in this ice is larger than Antarctica's sub-glacial lakes. The addition of hundreds of metres of ice at the base of an ice sheet bends the overlying ice and causes uplift of the surface of the glacier. This changes the slope and flow of the ice. The thickest sub-glacial ice was 1100 metres and this pushed the top of glaciers up 410 metres to reflect the shape of the added basal ice.

Antarctica has another little surprise. Underneath the ice sheets are volcanoes. The last big eruption was in Roman times and Mt Erebus is continually restless. Addition of heat from below could cause massive melting and detachment of a large block of ice. It's happened before and it will happen again.

As snow falls, it traps air. This air is preserved as the snow becomes an ice sheet. This trapped air remains trapped and uncontaminated in ice otherwise it cannot be used to measure past atmospheres. Antarctic ice core (Siple) shows that there were 330 parts per million of carbon dioxide in the air in 1900; Mauna Loa Hawaiian carbon dioxide measurements in 1960 show that the air then had 260 parts per million carbon dioxide. Either the ice core data is wrong, the Hawaiian carbon dioxide measurements are wrong or the atmospheric carbon dioxide content was decreasing during a period of industrialisation.

As in all other areas of science, uncertainty rules.

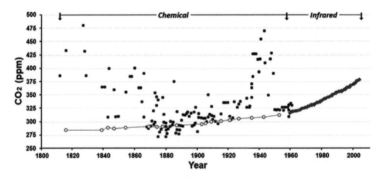

Figure 5: *Atmospheric carbon dioxide measurements by chemical (1812-1960 Europe, accuracy 1-3%, black dots) and infrared methods (1959-2009 Mauna Loa, accuracy <0.1%, dotted line) compared with ice core measurements (circles). Ice core measurements have been "adjusted" and, despite slightly lower accuracy, the historical chemical measurements show a huge variation in atmospheric carbon dioxide.*

CARBON DIOXIDE

Sea level

Sea level is not just a matter of how full the oceans are. The normal situation is that sea level rises when more water is added. However, sea level change is far more complicated than that. Although sea levels change with climate, the margins of continents where many people now live also rise and fall by tens of metres. The Earth actually bends very easily. Areas that were covered with kilometres of ice only 12,000 years ago were pushed down by the weight of ice and now are rising because the ice has melted long ago. The margins of continents appear to be continuously rising. As the ocean floor is pulled apart, the land on each side of the ocean rises. We see this with the coastal ranges on the east coast of Australia and the west coast of New Zealand. The ocean crust is thinner than the continental crust and loading the oceans with more water sometimes can make the ocean floor sink and continental land masses rise. It sounds bizarre but the addition of water to an ocean can make sea level fall.

Deltas such as the Mississippi Delta sink under the load of sediment because the sediment is compressing. In other deltas, the addition of sediment from the river system is greater than subsidence and so the delta grows. Bangladesh grows in area 20 square kilometres each year. Where the ocean floor is being pushed under a continent, the edge of the continent is also pulled down. However, the continent margin bends and after it can be pulled down no more, it will suddenly lift (and produce earthquakes and tsunamis). Such uplift changes both the land level and the sea level. This is what happened with the Japanese earthquake on 10th March 2011.

To argue that temperature and sea level are increasing depends on when measurements first started and depends on what is measured. Since 1842, it has been known that coral atolls rise with rising sea level. By ignoring such validated knowledge, scare campaigns about island states have gained some traction. In the Pacific Ocean, a recent study of 27 atolls shows that they either remained stable (11%) or increased in area (86%). Only 3% of islands underwent reduction in area. Long term studies of the Maldives shows that they have risen. Emotive scare campaigns about sea level are numerous. Images from the top of the Majuro atoll (Marshall Islands) accompanied by announcements that a one-metre sea level rise will erode 80% of the atoll make good television. Especially when given on location by the Australian government's parliamentary secretary for the Pacific, Richard Marles. Either the parliamentary secretary is ignorant or he is withholding vital scientific information. Research by Murray Ford (Hawaii University) showed an average sea level rise at the atoll of 3 millimetres per year, the atoll lagoon is decreasing and the ocean-facing shore of the atoll is growing. The sea level rise actually results from sinking due to land reclamation for residential, commercial and industrial development.

Since the zenith of the last glaciation 20,000 years ago, global sea level has risen 120 metres to a peak known as the Holocene Optimum when average sea level was 2 metres higher and global temperature about 2°C warmer than now. Sydney and Brisbane airports were tidal mangrove swamps and other examples are known from elsewhere. This is published research that the IPCC has swept under the carpet. Sea level has decreased since then at an average of 0.3 millimetres per annum but it goes up and down like a roller coaster. The temperature has declined in a similar roller coaster manner as well and the Minoan, Roman and Medieval warmings were the peaks that were warmer than today. It is currently rising but at a reduced rate. It is a very long bow to argue that human emissions of carbon dioxide create global warming which then results in melting of ice and expansion of seawater to give an increased sea level rise. During the 20th century when human emissions of carbon dioxide greatly increased, there were both warming and cooling periods.

Coastal planning based on global sea level rise is asinine because it ignores local compaction, sedimentation, uplift and subsidence. Not only does the sea level rise and fall, the land level also rises and falls.

Many of the rises and falls of both sea and land level are local, others can be regional and others can be global. Computer-modelled sea level projections by the IPCC and governments have already been shown to be hopelessly wrong. Since the release of the 2007 IPCC report, the rate of sea level rise predicted by the computer models has been shown by measurements to be 25% too high. Some 15% of the observed sea level rise comes from pumping of groundwater for human use into the oceans.

Areas covered with ice sheets during the last glaciation (116,000 to 10,500 years ago) sank. With the collapse of the ice sheets in the current interglacial, some lands are rising (e.g. Scandinavia, Scotland) and others are sinking (south-east England, The Netherlands). Oulo (Finland) is rising at 6.2 mm per year and this rise is used as a correction factor for IPCC sea level changes. Tide gauges show that Galveston (Texas) is sinking at 6.6 millimetres per year, which makes it look as if sea level is rising by 6.6 millimetres per year. The IPCC does not correct for this subsidence and only corrects for rising land that was pushed down by the weight of ice. Selectively correcting measurements makes it look as if sea level is rising quickly.

Tectonic processes change sea level. History shows us that some port cities (e.g. Ephesus, Turkey) are now inland as a result of land rise whereas other cities (e.g. Lydia, Turkey) are submerged as a result of subsidence. These two ancient cities are close to each other and suffer regular earthquakes that produce rises and falls of land. These areas were not covered by ice during the last glaciation. In both the Maldives and eastern Australia, relative sea level has fallen over the last 5,000 years. The Maldives is 70 centimetres higher now than in the 1970s probably due to the northwards movement of the Australian plate and eastern Australia is 2 metres higher than 5,000 years ago. Without a detailed knowledge of local land rises and falls, subsidence and sedimentation, sea level predictions are only speculation.

In the past 2,000 years, sea level has oscillated with five peaks reaching 0.6 to 1.2 metres above the present level. There is actually a Commission on Sea Level and Coastal Evolution, an independent body unrelated to the IPCC. This is a group that actually measures sea level and does not make computer predictions based on incomplete information. It states that by 2100 AD, sea level will have risen by 5 plus or minus 15 centimetres. This means that as a result of their comprehensive specialist studies, they cannot determine whether sea level will rise, fall or be static for the rest of the century as the uncertainty of their measurement is far greater than the actual measurement.

Governments may ask the simple question: How much do we need to spend to cope with sea level rise? This assumes that the sea level is rising in their area, which is commonly not the case. For example, how much should we spend to prevent $4 trillion of Florida real estate being destroyed in the next 500 years? At a rate of discount for the future of 6% per year, the answer is 15 cents! Economic theory tells us that we cannot control what may or may not happen in 500 years time. Common sense says: why bother? Geology says that there may actually be a sea level fall or a sea level rise. The level of uncertainty over a 500-year period is so high that we can equally as well argue that the climate may actually cool and sea level may drop.

The sea surface is not as level as one might expect. High and low pressure meteorological systems have an effect on sea level. There was a huge hump in the South Pacific Ocean. The bulge extending over an area larger than Australia from October 2009 to January 2010. Wind drove water into an area of an unusually stable high-pressure system. Off the coast of India, the sea level is considerably lower than the present global average because gravity is weak in one area. Near the coast of Greenland and the Andes Mountains, sea level is higher because of the gravitational pull of mountains. Ocean water sloshes between the east coast of Africa and the west coast of the Americas changing local sea levels.

Coastal areas are very sensitive to changes in climate and sea levels. In Western Australia during the last interglacial 125,000 years ago, there were staghorn coral fringing reefs at Fairbridge Bluff on Rottnest Island offshore from Perth. Sea level was 3 metres higher than at present and the seas were far warmer. There were no human industrial emissions of carbon dioxide 125,000 years ago. In the current interglacial, no such coral grows on Rottnest Island and the nearest staghorn corals are 500 kilometres north at Houtman Abrolhos. This suggests that the current interglacial is cooler than the previous interglacial. In polar regions, some suggest that during the last interglacial period, sea level was 4 to 9 metres higher than now and temperature was 3 to 5°C higher than now. Local conditions are such that there can be no definitive figure except that sea level was higher than now and it was warmer. Near King Sound (Derby, Western Australia), there are thickly vegetated sand dunes. These extend as fossil dunes beneath the coastal mud flats and show that dune formation took place 20,000 years ago at the peak of the last glaciation. Northern Western Australia was then arid, cold and windy, sea level was 100 metres lower than now and there was no vegetation on shifting dunes. Since then, the climate has become tropical, sea level has risen and tidal mudflats have covered what were treeless dunes.

Not all sea level change is due to climate. Some 97% of groundwater extracted for human use ultimately ends up in the oceans. It is thought that this may contribute up to 0.8 millimetres per year of sea level rise. Not all land level changes are natural. Extraction of groundwater, oil and gas results in compression of sediments and sedimentary rocks and earth tremors are commonly associated. As a result many places are sinking as a result of human activities (e.g. Bangkok, Mexico City, Venice 1920s to the 1970s). Vibration and the weight of cities also results in subsidence. The loading and unloading of dams with water also changes the land level and also produces faint earth tremors.

We are told that global sea levels have risen about 1.7 millimetres in the 20th century. This, it appears, is a result of us putting carbon dioxide into the atmosphere. This, we are told, warms the planet, expands the volume of the oceans, melts polar ice and adds water to the oceans. However, since the last glaciation 11,500 years ago, sea level has been rising and we cannot claim that the small amount of sea level rise in our own life is our fault and the rest was natural. There is enormous controversy in the scientific literature about the 120-metre rise between 12,000 and 6,000 years ago as models and calculations show that Greenland and Antarctic ice contributed very little to this rise. The question that has not been answered is: Which part of the recent sea level rise was from human activities and which part was from the normal post-glacial sea level rise? Some 200 years ago, convicts engraved tide marks on rocks in Tasmania. These show no change in sea level since then.

Tide gauge readings since the 1930s in the USA show no increase in the rate of sea level rise. Yet to be resolved is the disparity between tide gauges and recent measurements from satellites that show sea level has been rising. The sea level measurements by satellite shows a rising trend but this is not actually measured. The problem with satellites is that they don't actually measure sea level. They measure gravity and computers use the observations of gravity to calculate a theoretical even surface of the Earth and from this there are various ways computers calculate sea level. Very small changes in the computer code and database can produce apparently rising seas, falling seas or a static ocean.

In 2006 Australia's climate commissioner Tim Flannery stated:

> Picture an eight-storey building by a beach, then imagine waves lapping its roof. So anyone with a coastal view from their bedroom window or kitchen window is likely to lose their house as a result of that change.

Those on the waterfront who listened to the great guru Flannery must have been thinking about moving so as not to lose the principal family asset with future sea level rise. Flannery thinks up such frightening scare scenarios from his Coba Point home, which just happens to be at sea level. Why hasn't our climate commissioner moved to higher ground? Is it because he does not believe what he says? When I see Flannery move to higher ground, stop travelling in carbon dioxide emitting jets and vehicles, have meetings in Canberra bureaucrats' offices with no central heating, no lights and no air conditioning, stop using coal-generated electricity for broadcasting, live in one of the numerous caves at Coba Point with no electricity and live as he preaches, then I might actually seriously consider one of his numerous scary predictions. If coastal real estate loses value, can Flannery be sued? Does Flannery want us to lower our standard of living while he lives in the land of luxury? If the green movement uses Flannery as their poster boy, does this mean that the green movement is not concerned about truth and credibility? Maybe they are only concerned about controlling your life?

Who else has a waterfront home? Kevin Rudd? Yes. Al Gore? Yes. The climate industry seems to have a large mortgage on hypocrisy. The Greens Party politicians, Ross Garnaut and other climate industry leaders flit around business-class in planes pumping huge amounts of carbon dioxide into the atmosphere and tell us that we should reduce our emissions or carbon footprint (whatever that might be). The proponents of a "Carbon Tax" and all those in the climate industry that travel to international conferences and doomsday prediction meetings are too willing to tell us to reduce our standard of living.

Tim Flannery does not seem to believe what he preaches. Nor does the Australian government's climate change adviser Ross Garnaut. In the Senate Hansard of April 16th, 2009, Garnaut states:

> There is enormous uncertainty in the science. The uncertainty is in both directions. On the whole, the uncertainty adds to the case for strong and early mitigation.

This is the government's climate change adviser stating it as it is. There is huge uncertainty in "climate science". An uncharitable person might think: What would Garnaut know about science?

A little later (25th January 2010) in front of the annual conference of Supreme and Federal Court judges, Garnaut states:

> No, the science is not settled on all the dimensions of a complex natural system that are important to human society: science is never settled in an absolute sense, and in a complex system the detail will be adjusted continuously as more data becomes available.

This is the government's climate change adviser not speaking under oath, stating it as it is and knowing full well that he will not get away with intellectual bankruptcy in front of judges. All of a sudden, we learn that science is never settled. This is contrary to what Garnaut says in public.

On *Lateline* (17th March 2011), we learn that the science is now suddenly settled and that there is no uncertainty:

> *Jones:* On Monday [Tony Abbott] told a community forum the science is not settled, it's not proven that carbon dioxide is not quite the environmental villain that some people make it out to be. Doesn't that indicate that there certainly is a cloud on his mind over this issue at the very least?
>
> *Garnaut:* Oh well, he was on the ignorant side on Monday night, but not on Tuesday. The Tuesday statement was consistent with the science.

In this interview, rather than telling the viewers about uncertainty and the fact that science is not settled, he plays the political partisan and tries to take a cheap shot at the Leader of the Opposition prompted by a sympathetic ABC interviewer. Imagine if Ross Garnaut underwent robust cross-examination in front of one of the Supreme Court judges he addressed in 2010.

4

TEMPERATURE

How is global temperature measured?

Five organisations publish global temperature data. Two (Remote Sensing Systems [RSS] and the University of Alabama at Huntsville [UAH]) are satellite datasets and three are land-based sets (National Oceanic and Atmospheric Administration [NOAA], National Climate Data Center [NCDC], NASA's Goddard Institute of Space Studies [GISS] and the University of East Anglia's Climate Research Unit [CRU]). All land-based sets depend on data supplied by ground stations via NOAA. These data centres perform some final adjustments to data before the final analysis. Adjustments are common and poorly documented.

Temperature measurements of the surface of the Earth are compiled into what is known as the HadCRUT data. The data derives from 5° latitude x 5° longitude grid cells covering the Earth's surface. There are a total of 2592 cells. Cells that entirely cover land comprise just 21.7% of all cells and 18.1% of the total surface area and their data comes from observation stations. Sea grid cells comprise 50.6% of the total cells and 53.8% of the total area, and their data comes from measurements of sea surface temperature. Cells that cover a mixture of land and sea account for 27.7% of cells and 28.0% of area, and their data might be from observation stations or sea surface temperatures, depending on what's available and what's regarded as most reliable. The oceans cover 70% of the surface of the Earth. Both sea surface temperature and land temperature measurements have errors and the mixing data from two different measurement methods creates greater uncertainties. In any given month of measurements, coverage is around 80%, a slight fall from early 1980s levels. Data from the observation stations is commonly adjusted by the local meteorological authority. Data generally is not collected from more than 50% of the Earth's surface.

US climate research has received more than $73 billion in funding over the last two decades. And what do we have? Suspect data. In 1999, the NOAA administrator (Jane Lubchenko) paved the way for the politicisation of science by stating:

> Urgent and unprecedented environmental and social changes challenge scientists to define a new social contract....a commitment on the part of all scientists to devote their energies and talents to the most pressing problems of the day, in proportion to their importance, in exchange for public funding.

Those who promote a doom-and-gloom future in return for research funds are the very same people caught fiddling the data in the Climategate fraud. Since the 1960s, some 62% of temperature-measuring stations have been removed and most of these were in the colder remote, alpine, polar and rural regions where temperatures were lower. Most measuring stations are now near the sea or are at airports and cities which bias measurements because of urban and industrial heat. According to the calculations of Lubos Motl, 31% of the stations used by the UK's Hadley Centre Climate Research Unit (HadCRUT), the standard for surface temperatures show that temperature fell since 1979. This agrees with the temperature trends over the same period posted by John Christy and Roy Spencer (University of Alabama). If there has been any global warming, then it has not been uniform. The measurement of a global temperature is not simple. Historical instrumentally recorded temperatures exist only for 100 to 150 years in small areas of the world. From the 1950s to 1980s temperatures were measured in many more locations. Many measuring stations are no longer active. The main global surface temperature data set is managed by NOAA which states: "The period of record varies from station to station, with several thousand extending back to 1950 and several hundred being updated monthly."

There now have been so many stations shut down, especially in cooler high elevation, high latitude, rural and remote stations, that a significant warming bias has entered the overall record from land stations. This is the main source of data for global studies, including the data reported by the IPCC. Average surface air temperatures are calculated at a given station location based on the following procedure: record the minimum and maximum temperature for each day; calculate the average of the minimum and maximum; calculate the averages for the month from the daily data; and calculate the annual averages by averaging the monthly data. Various adjustments are also made, so it is not actually that simple. The IPCC uses data processed and adjusted by the Climatic Research Unit of the University of East Anglia which states:

> Over land regions of the world over 3000 monthly station temperature time series are used. Coverage is denser over the more populated parts of the world, particularly, the United States, southern Canada, Europe and Japan. Coverage is sparsest over the interior of the South American and African continents and over the Antarctic. The number of available stations was small during the 1850s, but increases to over 3000 stations during the 1951-90 period. For marine regions sea surface temperature (SST) measurements taken on board merchant and some naval vessels are used. As the majority come from the voluntary observing fleet, coverage is reduced away from the main shipping lanes and is minimal over the Southern Oceans.
>
> Stations on land are at different elevations, and different countries estimate average monthly temperatures using different methods and formulae. To avoid biases that could result from these problems, monthly average temperatures are reduced to anomalies from the period with best coverage (1961-90). For stations to be

used, an estimate of the base period average must be calculated. Because many stations do not have complete records for the 1961-90 period several methods have been developed to estimate 1961-90 averages from neighbouring records or using other sources of data. Over the oceans, where observations are generally made from mobile platforms, it is impossible to assemble long series of actual temperatures for fixed points. However it is possible to interpolate historical data to create spatially complete reference climatologies (averages for 1961-90) so that individual observations can be compared with a local normal for the given day of the year.

It is important to note that the CRU station data used by the IPCC is not publicly available, neither the raw data nor the adjusted data. Only the adjusted gridded data is available and so for global temperature data we must trust what the CRU at the University of East Anglia feeds us. Before Climategate I might have been nervous about their adjusted gridded data. But now that the keepers of the raw data have been shown to be frauds, we can only conclude that all the data used by the IPCC comes under a very big black cloud. The NASA Goddard Institute for Space Studies (GISS) is a major provider of climatic data in the US (with NOAA as the source for GISS). Some 69% of the GISS data comes from latitudes between 30 and 60 degrees north and almost half of those stations are located in the United States. This certainly does not cover the globe and is totally unrepresentative for any global picture. If these stations are valid, the calculations for the US should be more reliable than for any other area or for the globe as a whole. However, many recording stations were closed down in the late 1980s and early 1990s.

In addition, the locations of many stations were moved to escape urbanisation. Some stopped collecting data during periods of conflict. With the number of temperature measuring stations changing over time, the so-called "global" record is not really global. Land surface thermometers increased from coverage of 10% of the land in the 1880s to about 40% in the 1960s. Since then, the coverage has been decreasing. Coverage has been redefined as the "percent of hemispheric area located within 1200 kilometres of a reporting station." This means that in remote areas where there may be no stations in a 5x5 degree grid box, the temperature is estimated from the nearest station within 1200 kilometres. In my experience, there are considerable variations in temperature over 1200 kilometres, especially in remote or high latitude areas.

The closure of temperature measuring stations was not completely random. There has been a decrease in the number of Russian measuring stations and the Hadley Centre ignores many continuous long-term records. In Canada, the number of stations has decreased from more than 600 to 50. NOAA used only 35 of the Canadian stations. The percentage of those at lower elevation (less than 100 metres above sea level) tripled and those above 1000 metres reduced by half. More southerly locations using population centres hugging the US border dominated over northerly areas. In fact, only one thermometer remains in Canada for everything north of the 65th parallel. This site is called Eureka because life is more abundant and the summers are warmer than elsewhere in the High Arctic thereby creating a temperature bias for the whole region. Hourly readings from Russia and Canada can be found on the internet yet these are not included in the global data set.

China had a great increase in measuring stations from 1950 (100 stations) to 1960 (400) and then a great decrease to only 35 in 1990. Even Phil Jones was able to show that recorded temperatures in China rose due to increased urbanisation. In Europe, high mountain stations were closed and there were more measurements from coastal cities, especially along the Mediterranean Sea. In northern Europe, the average results showed warming as the number of stations decreased. After 1990, Belgium showed warming yet the adjacent country (the Netherlands) showed no warming. A similar story is apparent for South America where alpine stations disappeared and the number of coastal stations increased. There was a 50% decline in measuring stations.

African raw data also shows a warming bias. In North Africa, stations from the hotter Sahara were used in preference to those from the cooler Mediterranean and Atlantic coasts. In Australia and New Zealand, some 84% of stations are at airports and stations that closed were at cooler latitudes.

In the US, some 90% of all measuring stations did not appear in the global historical climate network (GHCN) version 2 climate models. Most stations remaining were at airports and most of the high mountain measuring stations have closed. In California, only San Francisco, Santa Maria, Los Angeles and San Diego were used. The warming trend is still not significant despite this bias. Warmist James Hansen states: "The US has warmed during the past century, but the warming hardly exceeds year-to-year variability. Indeed, in the US the warmest decade was the 1930s and the warmest year was 1934."

Any computer projections of global climate using land temperature measurements are deeply flawed and we should treat all claims, models, trends and predictions based on these with a very large pinch of salt.

The oceans play the dominant role in perpetuation and mediation of natural climate change as they contain far more heat than the atmosphere. Density variations linking the Northern and Southern Hemispheres of the Pacific and Atlantic Oceans via the Southern Ocean drive the ocean circulation system that controls hemispheric and global climate. Differences in temperature and salt concentrations produce these density variations. The oceans both moderate and intensify weather and decadal climate trends due to their great capacity to store solar heat. The process involves global currents, slow mixing, salt concentration variations, wind interactions and oscillations in heat distribution over very large volumes.

Ocean currents are global heat conveyor belts. The Northern Pacific Decadal Oscillation, the El Niño-La Nina Southern Pacific Oscillation, the long period Southern Pacific Oscillation, the Gulf Stream-Northern Atlantic Oscillation, the Indonesian Through Flow, the Indian Ocean's Agulhas Current, the Southern Ocean

Circumpolar Current and many other ocean currents and cycles have a large decadal-scale effect on weather, regional climate and global climate. In some parts of the global ocean heat conveyor, natural variations in heating, evaporation, freshwater input, atmospheric convection, surface winds and cloud cover can greatly influence ocean currents close to continents. In turn these modify carbon dioxide uptake or degassing, storms, tropical cyclone frequency, abundance of floating life, droughts and sea level changes. The ocean transfer of heat is one of the major driving force of climate. For scientists, especially oceanographers, to suggest that it is only human emissions of carbon dioxide that drive global warming exposes their non-scientific agenda.

The oceans occupy 71% of the Earth's surface. Sea surface temperature is measured. The Hadley Centre only trusts measurements from British merchant ships. These mainly use shipping routes in the Northern Hemisphere yet the Southern Hemisphere oceans occupy 80% of the ocean surface area. The change in methods over time from water collection in canvas buckets to later engine water intakes at various water depths has introduced measurement uncertainties. Ocean temperatures from ships, buoys and satellites also present opportunities for "adjustment", as the Climategate emails show. Satellite measurements were removed from public scrutiny by NOAA in July 2009 after complaints of a cold bias in the Southern Hemisphere. Immediately after that, both ocean and global temperatures mysteriously increased. The 3341 ARGO buoys have been measuring sea surface temperature since mid-2003. There has been a slight cooling trend. ARGO buoy data is not used in monthly assessments of global temperature.

What is unprecedented is that in the history of science so much data has been tampered with by so few.

Urban effects

Urban areas are warmer than surrounding rural areas. This is especially the case at night. Airports originally on the outskirts of urban areas have seen cities grow around them and temperatures rise. A very large number of measurements for global temperature studies are from airports. Long ago T. R. Oke created an equation for urban heat island warming.[6] A hamlet with a population of 10 has a warming bias of 0.73°C, a village with 100 people has a warming bias of 1.46°C, a town with a population of 1000 has a warming bias of 2.2°C and a city of a million people has a warming bias of 4.4°C.

The World Meteorological Organisation and NOAA supposedly have strict criteria for temperature measuring stations. Measurement stations should be located on flat ground surrounded by a clear surface and more than 100 metres away from local heat sources, tall trees, artificial heating or reflecting surfaces such as buildings, parking areas, roads and concrete surfaces. Temperature sensors should be shaded from direct sunlight, ventilated by the wind 1.5 metres above mown grass no higher than 10 centimetres. Anthony Watts found that 89% of the US measuring stations (i.e. more than 1,000 stations) did not meet the official standards for temperature measurement regarding the distance between stations and adjacent heat sources. Many received back reflection from concrete and buildings, most received heat from hot car and aeroplane exhaust fumes and many were sited next to air conditioning units that emitted heat. Watts concludes: "The raw data produced by the stations are not sufficiently accurate to use in scientific studies or as a basis for public policy decisions."

These biased measurements account for more than 50% of the measured warming since 1880. NOAA initially denied that it was an issue and then asked the government for $100 million to upgrade and correct the locations of 1,000 climate stations.

---

6 Urban heat-island warming= 0.317lnP, where P=population.

Adjusting measurements

After the raw temperature measurements are collected, warts and all, further adjustments are made. Curiously, each adjustment produces more warming. MIT meteorologist Richard Lindzen commented:

> When data conflicts with models, a small coterie of scientists can be counted on to modify the data to agree with models' projections.

By extracting old data from papers of James Hansen and comparing them with data downloaded from NASA's GISS site in 2007 and 2010, we can see a progressive global warming. This is certainly man-made, created by men making adjustments to the data, blending badly sited urban data with correctly-sited rural data and then, in 2007, removing the urban adjustment for US data sites. The frequency and direction of the NOAA adjustments to US data were increased in 2007 at the same time as inarguable satellite data showing global cooling became public knowledge.

NOAA's "homogenisation process" has been shown to significantly alter the trends in many stations where the location suggests the data is unreliable. In fact, adjustments account for virtually all the warming trend. Unadjusted data for the most reliable sites (i.e. rural sites) show cyclical multi-decadal variations and no long-term warming trend. Former NASA scientist Ed Long showed that after adjustment, the rural data trend agreed with the urban data trend when an artificial warming trend was introduced. Urban warming was allowed to remain in the urban data sets and a warming bias was artificially introduced to rural data sets that in their unadjusted state showed no warming.

Both NOAA and NASA resisted Freedom of Information requests for the release of all the unadjusted data and documentation for all the adjustments made. The US Data Quality Act requires that published data must be able to be replicated by independent audits. That is currently not possible given the resistance posed, despite promises of transparency.

Judith Curry (Georgia Tech University) comments:

In my opinion, there needs to be a new independent effort to produce a global historical surface temperature dataset that is transparent and that includes expertise in statistics and computational science…The public has lost confidence in the data sets…Some efforts are underway in the blogosphere to examine the historical land surface data (e.g. such as GHCN), but even the GHCN data base has numerous inadequacies.

The term global warming is based on an increasing trend in global average temperature over time. This is based on measurements. Or is it? The IPCC reported in 2007 in Chapter 3 that: "Global mean surface temperatures have risen by 0.74°C ± 0.18°C when estimated by a linear trend over the last 100 years (1906–2005)."

This is misleading because 100 years ago, thermometers did not have such a high order of accuracy.

The temperature record is not a good record and it is really only the long-term rural stations that provide meaningful raw data. Because of the closure of so many stations, by 2005 there were more measuring stations in the US than in the rest of the world. Since 2000, NASA has further "cleaned" the historical record by making adjustments for area, time of observation, equipment change, station history adjustment, fill missing data and urban warming adjustment. The rationale for the temperature station adjustments was to make the data more realistic for identifying temperature trends. The end result of all the adjustments is that additional apparent warming is created and one wonders whether this was accidental or deliberate. For example, the closest rural station to San Francisco (Davis) and the closest rural station to Seattle (Snoqualmie) have both had the older part of the record adjusted downwards. The end result of this is to give a trend that looks as if there has been more than a century of warming in rural areas when the raw data shows a very different story.

In Australia, there are a number of city-based measuring stations. Measurements are adjusted. Australia certainly has man-made global warming. It has nothing to do with carbon dioxide and occurs at the stroke of a pen. If all 328 stations are analysed from 1881 to 2008, the raw data shows a statistically insignificant warming of 0.07°C. This has to be looked at in context as many historical measurements have an uncertainty greater than 0.5°C.

Similar adjustments are made elsewhere. For example, the raw data from Melbourne show an increase in temperature from about 1950. This coincides with construction of large buildings thereby producing more back reflection, more vehicles and heating/cooling systems in office blocks. The raw Melbourne data show the urban heat island whereas the adjusted data show that Melbourne was warmer when there was little urban heat island effect and shows a warming from 1950. The raw data from Darwin show a slight cooling over the 150-year record. In the Second World War, the Darwin station was bombed and later moved from the city to the airport. The pre-1950 records have been adjusted downwards and so the adjusted record for Darwin looks as if there has been significant warming over the last 70 years. Exactly the same adjustments were made for Auckland with a rural station suffering substantial adjustments in the pre-1950 record such that the adjusted record looks as if there has been a sustained period of warming.

The National Institute of Water and Atmosphere (NIWA), a New Zealand government research organisation, records average NZ temperature for the past 100 years. This shows a warming since 1900. The NZ Science Coalition has obtained the raw data after long and painful political obfuscation by the Government and NIWA. The raw data, derived from seven long-term climate stations, have been adjusted. A record of the process for making adjustments just does not seem to exist. In fact, a couple of peer-reviewed papers show that the historical weather data show a trend of 0.3 plus or minus 0.3° C. This means that there was no change. This shows that there are no checks and balances operating for science funded by the public, that science has been politicised and that institutions such as universities, once regarded as the bastions of independent free thought and debate, are now promoting politicised and incorrect science.

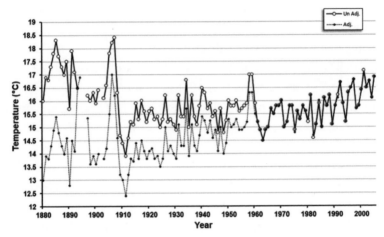

Figure 6: *Adjusted (thin line, dots) and raw (thick line, circles) temperatures for urban measuring station, Davis, Ca., USA. Note that adjustment gives apparent warming.*

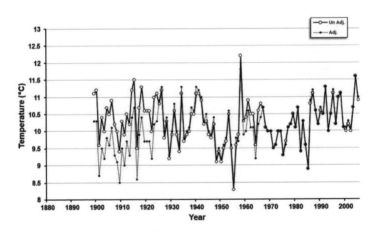

Figure 7: *Adjusted (thin line, dots) and raw (thick line, circles) for rural measuring station, Snoqualme Falls, Washington Sate, USA. Note that adjustment gives apparent warming.*

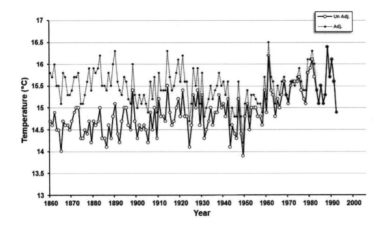

Figure 8: *Adjusted (thin line, dots) and raw (thick line, circles) temperatures for urban measuring station in the Melbourne CBD. The station location suggests influence by post-1960 increased traffic and buildings giving apparent warming.*

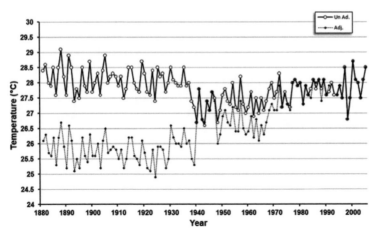

Figure 9: *Adjusted (thin line, dots) and raw (thick line, circles) for urban measuring station, Darwin, Australia. Note that adjustment gives apparent warming.*

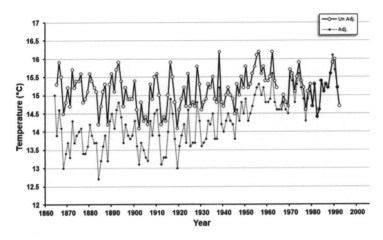

Figure 10: *Adjusted (thin line, dots) and raw (thick line, circles) for urban measuring station, Auckland, New Zealand. Note that adjustment gives apparent warming.*

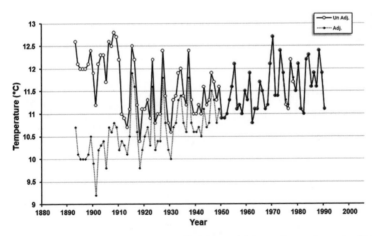

Figure 11: *Adjusted (thin line, dots) and raw (thick line, circles) for rural measuring station, Hokitika, New Zealand. Note that adjustment gives apparent warming.*

Since 1985, temperature records over land have been steadily drifting higher than sea surface temperatures and are now about 1°C higher. This may be the result of both closing rural (and often remote) land stations and the urban heat island effect. The US annual temperature from NOAA showed warming maxima in the 1930s and cooling minima peaked in the 1960s and 1970s. The NOAA web site had James Hansen stating:

> The US has warmed in the past century, but the warming hardly exceeds year-to-year variability. Indeed, in the US the warmest decade was in the 1930s and the warmest year was 1934.

There was constant friction as NOAA tried to infer warming from adjusted data. Indeed, by eliminating the urban heat island effect, warming since the 1930s suddenly appeared. Emails derived from Freedom of Information requests have revealed that David Easterling admitted:

> One other fly in the ointment we have is a new adjustment scheme for USHCN(V2) that appears to adjust out some, if not most, of the 'local' trend that includes land use change and urban warming.

Clearly the new computer models are creating pre-ordained conclusions for use by scientific activists. This is contrary to the 2009 conclusions of Brian Stone (Georgia Tech University) who found in 2009:

> Across the US as a whole, approximately 50% of the warming that has occurred is due to land use changes (usually in the form of clearing forest for crops or cities) rather than the emissions of greenhouse gases.

And:

> Most large US cities, including Atlanta, are warming more than twice the rate of the planet as a whole – a rate that is mostly attributable to land use change.

Hottest year on record

Every month, season and year, the world data centres release their assessment of the historic ranking for that period. NOAA announced that December 2009 was the eighth warmest December and the winter was the fifth warmest for the planet. No one believed NOAA. The Northern Hemisphere was then suffering its third consecutive brutal winter. Many places were the coldest for decades and some had the highest snowfall in history.

These shock-horror revelations that a particular year is the warmest on record implies that it is due to us humans emitting carbon dioxide and, as a result, the planet is getting warmer and warmer. But what does this really mean? We measure temperature by thermometer but sediments, fossil life, ice, caves and soils also provide us with a record of past temperatures. It was far warmer for most of the history of planet Earth than now. In the last interglacial some 120,000 years ago, it was warmer than now. Most of the past 10,500 years have been warmer than the present. With the exception of a few short sharp cold periods, Greenland temperatures were considerably warmer than now for at least the last 9,100 years. Around 6,000 years ago, it was far warmer than now.

It was announced on 28th December 2010 that 2010 was the hottest year since records have been kept. A decade ago it was the super El Niño year of 1998 but even this year trailed 1934 by 0.54°C. This was in the 1910-1940 warming period. Since then, NASA has "adjusted" the US data for 1934 downwards and the 1998 data upwards and just forgot to tell us about the strong El Niño in early 2010. This now means that 2010 was hotter than 1998, which was hotter than 1934 making it look as if temperature is increasing. Regardless of which year wins the temperature adjustment battle, how significant is it that we have warm temperatures after a 330-year period of gradual warming? Despite the cooking of the books, 2010 was only the 9,099th hottest year in our current 10,500-year long interglacial.

To support the story that 2010 was the hottest year on record, we were also told in December 2010, that there was less Arctic ice than ever before. What we were not told was that the loss of Arctic sea ice was within the range of variability measured from over 30 years of satellite measurements and that there was an equally rapid gain of Antarctic sea ice.

Even the Minoan, Roman and Medieval Warmings were warmer than now and people did not fry-and-die. And on a geological time scale, was 2010 a hot year? Definitely not. To make matters worse, the Australian media was in a frenzy reporting Hansen's declaration that 2010 was the hottest year on record (contrary to his earlier statements). They did not know that Hansen's year was the meteorological year that ended on 30th November and not a calendar year and hence he managed to avoid the cool Southern Hemisphere summer and the bitterly cold Northern Hemisphere December. Furthermore, Hansen compared 2010 to 2005 by using GISS records but did not state that these records have been modified so much that they differ greatly from the available raw data. Even forgetting the quibbles about adjusting data, the warmest year since 2000 has only exceeded the coldest year by 0.12°C. I am sure that 0.12°C is not life threatening and that we might just be able to adapt to such a change.

NASA's climate group is run by James Hansen, an outspoken environmental activist. He exaggerates and has been caught out in public many times. It is strange that the areas of the world where NASA detected the most significant warming in 2010 were also the areas where there were no weather stations. Yes, you read it correctly!

Many areas of the world have no historical temperature measurements (e.g. Third World countries). It was claimed that these areas showed great warming in 2010. It appeared that Greenland was much warmer but most of the 5 x 5 degree grid boxes in Greenland have no measuring stations and most of the other grid boxes have only one station. The two hottest areas of Greenland shown by

NOAA showed 5°C warming yet had no measuring station. Would it be wrong to state that these numbers were just plucked out of the air and, to our great surprise, showed significant warming that could not be validated by measurement?

It is the same for Siberia. In the 1940s, Siberia had areas that showed differences in warming and cooling by a few degrees. The measuring stations have now been closed down. Siberia appears now to have warmed in 5 x 5 degree grids where there are no measuring stations. As the Siberian historical data only goes back to the 1940s, it is hard to calculate a trend of warming or cooling. It appears that the Arctic of Canada is also sweltering. Almost all 5 x 5 degree grids have at least one measuring station. The only grid without a measuring station just happened to have warming of 5°C. Many stations are no longer maintained in the GHCN or CRUTEM3 databases yet NOAA claims that there has been a 4°C warming over the last 40 years. The record shows that similar warming occurred in the 1930s.

Temperature changes in the Canadian Arctic correlate with the El Niño-Southern Oscillation. The HadCRU data shows 60-year cycles of temperature, probably related to the El Niño-Southern Oscillation. As temperature has been increasing for 330 years since the Maunder Minimum at the end of the Little Ice Age, the cycles are secondary to the main trend. These 60-year cycles have been measured as far back as 2,000 years.

However, the climate industry says that there is a warming trend due to our emissions of carbon dioxide. Well, trends are trends, as the Whitehall ditty explains:

"A trend is a trend,
But when will it end?
Will it reach for the sky,
Or curl up and die?
Or will it just go around the bend?"

In summary, the surface temperature records in the pre-satellite era (1850 to 1980) have been so widely, systematically and uni-directionally tampered with that it cannot be claimed that there is human-induced global warming in the 20th century. These changes to the historical records mask cyclical changes that can be explained by oceanic and solar cycles. To increase the apparent trend of warming, earlier records of warming have been adjusted downwards. It is probable that these temperature databases are now useless for determining long-term trends. Furthermore, more than three quarters of the 6,000 stations that once reported are no longer being used in data trend analysis and some 40% of stations now report months when no data are available. These months where no data are reported required "infilling" which only adds to the uncertainty. The exaggeration of long-term warning by 30-50% is probably because of urbanisation, changes in land use, incorrect positioning and inadequately calibrated instrument upgrades. When changes in data sets were introduced in 2009, there was a sudden warming. However, satellite temperature monitoring has given an alternative to land-based stations and this record is increasingly diverging from the land-based measurements consistent with adding a warming bias to the land records.

When interviewed by the BBC, Phil Jones stated:

> Surface temperature data are in such disarray they probably cannot be verified or replicated.

If you cannot replicate the data, no matter how much it has been tampered with, then it is useless. No conclusions or predictions can be made.

You heard it from the horse's mouth.

5

HOW TO GET EXPELLED FROM SCHOOL

Background

Don't get too excited. Just in case you turned straight to this chapter, you first need a bit of background information. What is often done in science is to examine a hypothesis. The hypothesis underpinning human-induced global warming (and a 'Carbon Tax') is that human emissions of carbon dioxide drive global warming and that this warming may be irreversible and catastrophic. Following on from this is the catch cry: 'We must do something'.

What if doing something makes your life worse? The solutions to the human-induced global warming fad will all impact on every aspect of your life. There will be less energy available and it will be more expensive. Every consumer item will cost more. Do you really want to give up your mobile phones, iPods, iPads, DVDs, television, parties, travel, food and way of life to save the planet? Do people who claim they want to save the environment want to lead by example and have no hot water, no warm food, no heating, no air conditioning, no car transport and no modern life? Do people really want to make a number of trips a day to collect water from a well for the family in order to live a sustainable life? If young folk want to lead by example to save the planet, they should stop travelling, driving, drinking and eating. In essence, the easiest way to lead by example is to drop dead. Do school pupils really want to have the frugal life of their great grandparents?

169

Life is far better than 100 years ago. We eat better, live longer, have better housing and have a richer life. Environmental ideologies are attractive and form part of personal growth. But, an ideology embraced without analysis of practical aspects is vacuous. Global warming is a fad. Once there are consequences that affect a comfortable life, then another issue will be found. And embraced again with passion. What is the next scare campaign? Ocean acidification? Biodiversity?

Climate change has been with us for the 4,500 million year history of planet Earth. This is what climate does. It always changes. Changes in our lifetime may be natural. The Sun warms the Earth. Without the Earth being the right distance from the Sun, we would either freeze or fry. The amount of energy released by the Sun varies. The magnetic field of the Sun changes. This means that the amount of energy reaching the surface of the Earth also changes. We know that the moon and other planets have global warming and global cooling driven by the Sun. Unless there are cars, power stations and factories on other planets, global warming on other planets cannot be due to human emissions of carbon dioxide. A relatively inactive Sun allows cosmic radiation to form more low-level clouds and the Earth cools. The Sun heats the tropics the most and the poles the least and seasons come and go as the Earth orbits on its tilted axis. Many factors interacting on a variety of time scales drive the amount of solar heat reaching the Earth. The main factors are the ever-changing distance between the Earth and the Sun giving cyclical climate changes, the variable amount of energy emitted by the Sun and the variable amount of cosmic radiation reaching the Earth's surface. There are changes in the spin axis of the Earth. These produce a change in the length of the day and clocks need to be occasionally readjusted. Once there is a change in the spin of the Earth, then the ocean currents change.

For most of time, the planet's atmosphere had a far greater carbon dioxide content than now yet the planet both warmed and cooled. This clearly shows that other great forces drive climate change, that carbon dioxide does not and that the hypothesis is wrong. For more than 80% of the Earth's lifetime, our planet has been a warm wet greenhouse planet. The rest of the time, we have been in the deathly grip of ice ages. There have been six of these and during each there were glaciations (when ice sheets expand and sea level drops) and interglacials (when ice sheets retreat or disappear and sea level rises). During ice ages (and glaciations), the number of species of complex life on our planet is reduced. Each of these six major ice ages, including the current ice age that started 34 million years ago, commenced when the atmospheric carbon dioxide content was higher than now. If a high atmospheric carbon dioxide content is meant to drive global warming, then something went wrong. Six times. Clearly, other factors drive warming and cooling and carbon dioxide has not been a major force in the past. Why should it be a major force today, especially when the atmospheric carbon dioxide content is relatively low. These six great ice ages show that the hypothesis is wrong.

There have been times in the geological past when there was a massive injection of carbon dioxide into the atmosphere. At these times vegetation thrived. These periods (Carboniferous, Cretaceous, Tertiary) saw massive volcanic activity which changed the Earth. Volcanoes emitted monstrous amounts of carbon dioxide and volcanic heat may have released methane from sediments and rocks. The atmosphere had more carbon dioxide in it than now and the planet was warmer. Instead of putting stress on life, it thrived in these times of high carbon dioxide. The overall change in temperature was less than temperature changes occurring in a day. Despite having far more carbon dioxide in the atmosphere than humans could emit from burning all fossil fuels, there was no irreversible runaway greenhouse, the warmer climate was not forever and the planet went back to doing what the planet does. Storing carbon dioxide in life, soil and rocks. Again, this shows that the hypothesis is wrong.

Is climate change normal?

Ice core measurements show there were great temperature fluctuations between glaciation and interglacials. Snow trapped air, this air ends up trapped in ice at the poles and it can be extracted for measurement of the past atmospheric carbon dioxide content. Various chemical tricks can be used with ice to calculate the air temperature when the snow fell. Ash layers in the ice from volcanic eruptions can be used to measure time. Ice core measurements show that temperature rises some 800 to 2,000 years before the atmospheric carbon dioxide content rises. All this is telling us that, although carbon dioxide is a trace gas in the atmosphere and a minor greenhouse gas, it does not drive and has not driven global warming. Again, the hypothesis is shown to be wrong. It is clearly natural warming that drives emissions of carbon dioxide, mainly from the oceans.

This raises another question. If our planet has been warming for the last 330 years and there was a warming event 800 years ago, then the increasing amounts of carbon dioxide in the air may be because of degassing of carbon dioxide from oceans heated during the Medieval Warming combined with carbon dioxide emissions from humans. Temperature changes recorded in ice core from the Greenland Ice Sheet show that the global warming experienced during the past century pales into insignificance when compared to the magnitude of huge climate changes over the past 25,000 years. These temperature changes are really only valid for Greenland but do reflect what was happening in the Northern Hemisphere. During the peak of the glaciations, ice sheets thousands of metres thick covered North America, northern Europe and northern Russia and Alpine glaciers advanced rapidly.

Of course climate change is normal. It would be really weird would if there was no climate change on Earth. Ever since humans evolved, there have been huge rapid climate changes far greater than anything measured today or anything speculated about in the future. There has been supposedly a 0.8°C temperature rise over the last 150

172

years and the most scary predictions are for a 2 to 5°C rise over the next century. We can easily test the hypothesis that human emissions of carbon dioxide drive global warming and that this warming may be irreversible and catastrophic. Over the last 120,000 years there have been at least 25 periods of global warming when temperature rose by more than 8°C. These temperature rises took place during the latest glaciation and could not possibly be related to human emissions of carbon dioxide as there was no industry at that time. Unless natural processes drove atmospheric carbon dioxide up and down like a yo-yo, these 25 periods of massive warming again show that the hypothesis is wrong.

Over the last 25,000 years, at least three warming events were 20 to 24 times the magnitude of warming over the past century and four were 6 to 9 times the magnitude of warming over the past century. Again, this shows that the hypothesis is wrong. Some of the dramatic recent climate changes are well documented in history and validated in the scientific literature. These put the very slight late 20th century warming into perspective.

Some 24,000 years ago, during the peak of the last glaciation, there was a sudden warming of about 15°C. Now that is global warming! Was this warming due to smoke stacks emitting carbon dioxide? Of course not. The hypothesis is again wrong. Shortly after, temperatures dropped suddenly by about 7°C. Temperatures remained cold for several thousand years and fluctuated between 3°C warmer and 3°C colder than now.

About 15,000 years ago, there was another sudden rapid warming of about 12°C. This is far greater than anything measured today or predicted by the most scary catastrophist for the future. Was this warming due to heavy industry emitting carbon dioxide? Of course not. Yet again, the hypothesis is wrong. The large ice sheets that covered Canada and the northern USA, all of Scandinavia and much of northern Europe and Russia started to melt quickly and glaciers retreated.

173

A few centuries later, temperatures again suddenly fell about 11°C and glaciers re-advanced. Was this due to a sudden loss of carbon dioxide from the atmosphere? No. About 14,000 years ago, global temperatures again rose rapidly by about 4.5°C and glaciers receded. Was this due to human emissions of carbon dioxide? No. Again, the hypothesis is wrong.

About 13,400 years ago, global temperatures dropped, this time by 8°C and glaciers re-advanced. Was this due to a loss of carbon dioxide from the atmosphere or other factors such as the Sun, ocean heat changes and ice sheet collapse? About 13,200 years ago, global temperatures again increased rapidly by 5°C and glaciers re-advanced. Was this due to coal-fired power stations releasing carbon dioxide? No, they have only been with us for just over 100 years. Again, the hypothesis is wrong.

Some 12,700 years ago temperatures fell quickly by 8°C and a 1,300-year cold period, the Younger Dryas, began. After this 1,300-year period of intense cold, global temperatures rose very rapidly by about 12°C marking the end of the Younger Dryas cold period and the end of the latest glaciation. Did this glaciation end rapidly because we sinful humans suddenly put carbon dioxide into the atmosphere? Of course not, there was no industry and no agriculture then and again the hypothesis is wrong. Between the end of the Younger Dryas and 6,000 years ago, there was a prolonged period of warming with a 500-year period of cold. Some 8,200 years ago, the interglacial warming was interrupted by a period of sudden global cooling. During this time, alpine glaciers advanced. The warming that followed the cool period was also abrupt. Neither the abrupt climatic cooling nor the warming that followed was preceded by atmospheric carbon dioxide changes and we can only conclude that again the hypothesis is wrong. After this short sharp cold snap, warming continued until 6,000 years ago. It was warmer then than now. What is now the Sahara Desert was grassland with patches of trees. Sea level in eastern Australia was about two metres higher than now. As there was no industry then pumping out carbon dioxide into the atmosphere, it can only be concluded that the maximum warmth in post-glacial times was natural. Again, the hypothesis is wrong.

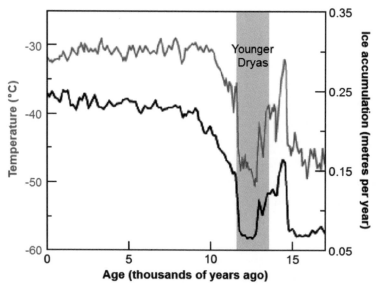

Figure 12: *Rapid post-glacial climate change as represented by the Younger Dryas. Both the magnitude and rate of climate change were far faster than any modern changes.*

There were again periods of warming and cooling in Egyptian and Minoan times. You are getting the story by now. These were natural and had nothing to do with human emissions of carbon dioxide. Again, history shows that the hypothesis is wrong. How many times does a hypothesis have to be shown to be wrong before it is rejected? Only once! The climate industry still clings to the carbon dioxide hypothesis and the only way this can be done is to ignore the history of the planet and pretend that the planet had a stable benign climate until the Industrial Revolution. This is not science. I would like to know what particular drug these catastrophists are 'on' as it clearly obliterates the past and creates a wonderful false reality!

Prior to the founding of the Roman Empire, Egyptian records show a cool climatic period from about 750 to 450 BC. If science is not your thing, then history has some good examples of how climate has changed quickly and naturally. Hannibal was able to take his army and elephants across the Alps in the winter of 200 BC. This would not be possible now. Julius Caesar conquered Gaul in about 50 BC. He had to build a bridge across the Rhine River (which now separates France from Germany) in order to wage war. The Rhine acted as a natural protective barrier for nearly 500 years but, as the Roman Empire was disintegrating in Gaul in the 5th century AD, the Vandals were able to walk across the frozen Rhine River to wage war. The Rhine has not been frozen in modern times.

The Romans wrote that the Tiber River froze and snow remained on the ground for longer periods. This does not now happen in Rome. During the Roman Warming after 100 BC, the Romans wrote of grapes and olives growing much farther north in Italy than had been previously possible because there was little snow and ice. Was this warming due to carbon dioxide released from fermentation of grapes to produce wine for the Romans? Hardly. Wine grapes were grown in northern England where it is now too cold to grow wine grapes. In 29 AD, the Nile River froze. It has not frozen since. Sea level did not rise during the 600-year long Roman Warming. Although the Romans did emit some carbon dioxide from smelting and agriculture, the amounts emitted were tiny compared to now. In Roman times it was warmer than now but this could not be due to human emissions of carbon dioxide and again the hypothesis is wrong.

During the Dark Ages, it was very cold. A strange event occurred in 540 AD when it became even colder. Trees grew very slowly, the Sun appeared dimmed for more than a year and temperatures dropped in Ireland, Great Britain, Siberia, and North and South America. In 800 AD, the Black Sea froze. It has not frozen since. If carbon dioxide drives warming, then was there a sudden event when carbon dioxide was reduced? No. All this shows that warmings and coolings during human history are natural and normal. Clearly the hypothesis is wrong again.

In the Medieval Warm Period (900 to 1300 AD) in Europe, grain crops flourished, the tree line in the Alps rose, the population more than doubled and many new cities arose. Great wealth was generated, as always happens in warmer times. The warmer climate allowed the Vikings to colonise Greenland in 985 AD. Ice-free oceans enabled Viking sea travel as far as Canada and the fishing grounds were enlarged. The Vikings grew barley and wheat in places that are now snow-covered and the depths of the graves show that there was no permafrost. In France and Germany, grapes were grown some 500 kilometres north of the present vineyards, again showing that it was far warmer than now. In Germany, where grapes are now grown at a maximum altitude of 560 metres, the Medieval vineyards were up to 780 metres altitude showing that temperature was warmer by about 1.0-1.4°C. Wheat and oats were grown at Trondheim (northern Norway), again suggesting climates 1°C warmer than now. Prolonged droughts affected southwestern USA and Alaska warmed. The Medieval Warming had a considerable effect on South America. Lake sediments in central Japan record warmer temperatures, sea surface temperatures in the Sargasso Sea were approximately 1°C warmer than today and the climate in equatorial east Africa was drier from 1000 to 1270 AD. Ice core from the eastern Antarctic Peninsula shows warmer temperatures during this period. The Medieval Warming was just not some local warming event in Europe. It was global.

In modern times, the Modern Warming has been no greater than 0.8°C. This we are told is due to human activities yet the Medieval Warming of almost twice this amount was natural. The obvious question arises: Which part of the Modern Warming is natural and which part is due to humans? The Medieval Warm Period was warmer than now, was global and clearly was not driven by power stations and factories emitting carbon dioxide. Sea level did not rise during this 400-year period of warming. The Medieval Warming shows again

177

that the hypothesis is wrong. And what is this hypothesis again? We are testing whether human emissions of carbon dioxide drive a global warming which may be catastrophic and irreversible.

The Little Ice Age (1300 to 1680 AD) started with a rapid decrease in temperature of 4°C. Glaciers expanded worldwide. This was not due to a sudden decrease in atmospheric carbon dioxide. Greenland glaciers advanced and pack ice expanded. The grain-dependent population of Europe suddenly faced a cold and variable climate with early snows, violent storms, catastrophic flooding and massive soil erosion There were constant crop failures, livestock died and famine became a great killer. Three years of torrential rains led to the Great Famine of 1315-1317 AD and the population was further reduced by the spread of the Black Death plague. Wine production decreased, the growing season for all crops was shortened, rivers such as the Thames in London and canals in the Netherlands froze. All travel became hazardous. Glaciers advanced in the Alps, Papua New Guinea, the Andes and New Zealand and they buried villages in the Swiss and Austrian Alps. When New York Harbour froze in the winter of 1780 AD, people could walk from Manhattan to Staten Island. In June 1644, the Bishop of Geneva took his flock to pray on a glacier that was advancing "a musket shot" every day. Even prayers to the Almighty did not change climate and it is extraordinary human arrogance today to think that if we humans twiddle the dials, then we can change climate. In the Little Ice Age, sea ice surrounding Iceland extended for great distances in every direction, closing many harbours and preventing fishing trips. The population of Iceland halved and the Viking colonies in Greenland died out in the 1400s because they could no longer grow enough food or get through the ice for fishing. In parts of China, warm weather crops that had been grown for centuries were abandoned. In North America, early European settlers experienced exceptionally severe winters. If carbon dioxide is the cause of global warming, then the past global warmings should mirror the rise in carbon dioxide. They don't. The Little Ice Age again shows the hypothesis is wrong.

We know from the Roman Warming, Dark Ages, Medieval Warming and the Little Ice Age and the last glaciation that climate changed very rapidly and that climate changes are not driven by the rise and fall of carbon dioxide. Over the last 500 years, the ice cores from Greenland, the position of Northern Hemisphere glaciers and lake sediments show that temperature oscillated about 40 times with changes on average, every 27 years. None of these changes could possibly have been driven by changes in atmospheric carbon dioxide or human emissions of carbon dioxide.

Widespread thermometer measurements started to be collected in the mid-19th century although there is a longer thermometer record in central England. Early measurements show a warming from 1860-1880, cooling from 1880 to 1910, warming from 1910 to 1940, cooling from 1940 to 1977, warming from 1977 to 1998 and then cooling from 1998 to the present. In the 21st century, temperatures are falling yet the amount of carbon dioxide being emitted to the atmosphere is increasing. Some have argued that this falling of temperature is due to increased particles in the atmosphere emitted from China. In the coolings of 1880-1910 and 1940-1977, there is no indication that they were due to increased particles in the air from China or anywhere else. Some have argued that increased particles in the air from World War II (1939-1945) drove the cooling from 1940-1977. However, World War I (1914-1918) was in a period of warming. If this speculation about dust from war is correct, there should have been cooling in World War 1. There was not. These crazier and crazier explanations arise because those trying to argue that humans drive climate change want to make all climate changes since the Industrial Revolution in the 19th century of human origin. Why is it that climate change by natural processes suddenly stopped because we had an industrial revolution? If we look at the percentage of human emissions of carbon dioxide from the 19th century to 2010, then the great increase in emissions took place over the last 60 years.

Year	Percentage of all human CO_2 emissions (to 2010)
1850	< 1%
1910	5%
1945	15%
1963	25%
1984	50%
1998	75%
2010	100%

Table 1: *Human emission of carbon dioxide since the Industrial Revolution.*

This table shows that the warmings of 1860 to 1880 and 1910 to 1940 could not be driven by massive emissions of carbon dioxide by humans. They can only be natural. Even as the emissions of carbon dioxide were increasing, there were events of cooling (1880-1910, 1940-1977). If the hypothesis is correct, then there should have been no cooling periods. Clearly, the hypothesis is wrong. Yet again. If the warming of 1977-1998 was driven by a large increase in emissions of carbon dioxide by humans, then for the hypothesis to be correct, the rate of warming should have increased. It did not. The three temperature increases over the last 150 years have all been at the same rate. If increasing human additions of carbon dioxide to the atmosphere drive warming, then the rate of warming would have increased over time. The hypothesis is wrong. Some 25% of all human emissions of carbon dioxide occurred in the last 12 years. This has been a time of cooling, the opposite of what the hypothesis predicts. Again the hypothesis is wrong.

In 1940, human emissions of carbon dioxide began to rise sharply. By 1977, atmospheric carbon dioxide had risen yet temperatures fell about 0.5°C in the Northern Hemisphere and about 0.2°C globally. Why did the temperatures fall for more than 30 years (1940 to 1977) when carbon dioxide was sharply accelerating? Global temperature suddenly rose again during the Great Climate Shift of 1977 when the Pacific Ocean switched from its cool mode to its warm mode with no change in the rate of carbon dioxide increase. The 1977 to 1998 warm cycle ended in 1999 and a new cool period began. Therefore, rising carbon dioxide is not the major cause of global warming and there is some other process was driving temperature change. Again, this shows that the hypothesis is wrong. Temperature patterns since the Little Ice Age (1300 to 1680) show a very similar pattern; 25 to 30 year-long periods of alternating warmer and cooler temperatures during an overall warming from the Little Ice Age. These temperature fluctuations in the Little Ice Age took place before humans started to emit increasing amounts of carbon dioxide into the atmosphere. The magnitude of the only modern warming that might possibly have been caused by carbon dioxide (1978-1998) is insignificant compared to the earlier periods of warming. The sudden warming about 15,000 years ago caused massive melting of the ice sheets and this rate was not unprecedented because this is what happens at the end of every glaciation. The rate then was far higher than anything we see today.

The data for the 21st century shows that even though carbon dioxide emissions are increasing, temperature is not. It is clear that carbon dioxide cannot be driving temperature change. This proves that the hypothesis is wrong. If a scientific hypothesis is wrong just once, then it has to be abandoned. Geology and history shows that the hypothesis is wrong time and time again. The climate industry continues to embrace a hypothesis that is wrong? Why? Because we are not dealing with science.

Carbon dioxide is a greenhouse gas, atmospheric carbon dioxide is increasing and every molecule of carbon dioxide humans emit causes a decreasing and very slight amount of warming. Surely that is proof that human emissions create global warming? No. Carbon dioxide is a trace gas and something else is causing warming. The world has been warming about half a degree per century since the depths of the Little Ice Age in 1680. During this time, the Maunder Minimum, there was no sunspot activity. The warming trend was as strong in the 1700s and 1800s as it was in the 1900s. This warming commenced well before there were large emissions of carbon dioxide by humans. Within this longer term warming trend since 1680 there is a pattern of 25-30 years of warming followed by 25-30 years of cooling. Despite the billions of dollars that have been spent looking for empirical evidence of man-made global warming, nothing has been found. If it were found, we certainly would have heard about it. Instead, it is claimed that the evidence for human-induced global warming is hidden in climate models, many of which contain assumptions and guesses about how the world works.

The 20th century was a time of increasing and variable solar activity and it appears that solar activity is declining. We have also commenced another 25 to 30 year period of cooling. If human-induced global warming in the last 30 years of the 20th century was as we are led to believe by the climate industry, then there is only one conclusion: nonsense. If so many more intense periods of warming occurred naturally in the past without any increase in carbon dioxide, why should the mere coincidence of a small period of slight warming in the late 20th century be blamed on carbon dioxide? Nature just refuses to co-operate with the catastrophist dogma because temperature has risen and fallen far quicker in the past than in modern times. The influence of human additions of carbon dioxide to the atmosphere is so small that the effect cannot be measured. You can see the problem the climate catastrophists have. This is why they ignore history and ignore geology and don't want to debate natural scientists. In order to believe human-induced global warming, propaganda requires the climate industry to ignore simple and overwhelming contrary evidence and to revile those that present this contrary evidence.

The IPCC stated that 97% of Earth's annual production of carbon dioxide is from nature and 3% from humans. Why is it that the human production of carbon dioxide is dangerous yet nature's is not? This question has never been answered by the climate catastrophists. Within the bounds of error, the average global temperature has not increased since 1995, temperature has actually been falling and over the last ten years atmospheric carbon dioxide levels have increased by 5%. Go figure. The elephant in the room is the evidence showing that more carbon dioxide in the atmosphere does not cause global warming despite carbon dioxide being a greenhouse gas.

6

ONE HUNDRED AND ONE QUESTIONS FOR YOUR TEACHERS

Maybe you should not try to ask all these questions yourself. Get some of your friends to help. Try these questions for a start:

1. Is climate change normal?

Yes. If the answer is not a very definite yes and there is diversion, discussions and exceptions, then your teacher is ignoring the past and is using classes for political purposes. No scientist denies that climate change is normal. If the answer is no, then your teacher does not know the basics and can not prepare you for life.

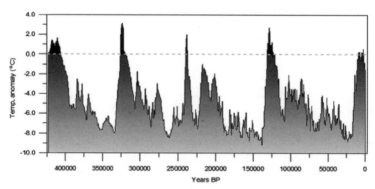

Figure 13: *Reconstruction of temperature from Vostok (Antarctica) ice cores showing that the current interglacial is not as warm as past interglacials. Climate changed on 100,000-year cycles with 90,000 years of glaciation and 10,000 years of interglacials. Variations in temperature and rates of warming during glaciation were far greater than any modern variation in temperature. Humans and other mammals (e.g. polar bears) survived both previous interglacials and glaciations. BP = Before the present.*

2. Is the global warming measured today unusual?

No. Modern warming from 1977 to 1998 is well within the range of variations seen in the past. If your teacher beats around the bush and does not give you a very definite no, then you know you are being taught propaganda. If the answer is yes, then you have proof that your teacher knows nothing of history and geology.

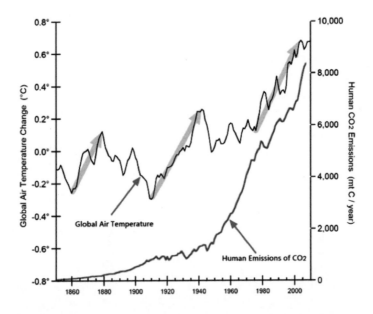

Figure 14. *Post-Little Ice Age thermometer temperature measurements showing three events of warming (1860-1880; 1910-1940; 1977-1998) when the rate of warming was constant. If the increased human emissions of carbon dioxide (CO₂) influenced temperature, then the rate of temperature increase from 1977 to 1998 would have been greater than the rates of previous temperature increase.*

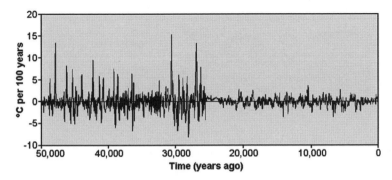

Figure 15. *Temperature change per century at the closing stage of the last glaciation derived from GISP2 ice core measurements. Rapid temperature change and great variability occurred during cold times rather than warm times and modern variability is well within the variability.*

3. What drove climate change before humans were on Earth?

It is the simple questions that always create grief. A great list of factors may be mentioned as written about earlier. However, past great climate changes were not driven by carbon dioxide.

4. Are we in a period of global warming?

The answer is both yes and no. There has been an overall warming for the last 330 years. Over the last century we have been warming and cooling and, over the last decade, there has been slight cooling. Considering longer time scales, we are currently in an interglacial and, on an even longer time scale, we have been in an ice age for the last 34 million years. In fact, planet Earth has been cooling for 50 million years. In the current ice age there have been large cycles of cooling and warming and shorter cycles of very slight cooling and slight warming. We are currently in one of those very slight cooling periods.

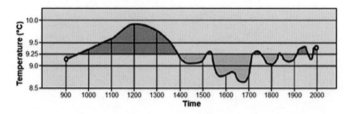

Figure 16: *The IPCC plot of temperature of climate change in Europe over the last 1,000 years showing the Medieval Warming (900 to 1400 AD), the Little Ice Age (1400 to 1680 AD) and the 330-year period of warming since the coldest time of the Little Ice Age. The rate of warming and the temperature in the Modern Warming (1680 to the present) is lower than the Medieval Warming. If human emissions of carbon dioxide created the Modern Warming, then the Medieval Warming and Little Ice Age would have to be expunged from the record. This is exactly what Mann and the IPCC did in the figure below.*

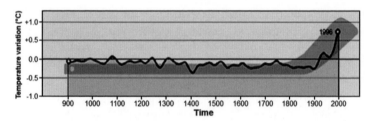

Figure 17: *The fraudulent "hockey stick" of Mann showing the temperature variation from 900 to 1996 AD based on selective use of tree ring measurements and omission of volumes of data.*

5. Will the 0.5°C warming we experienced from 1977 to 1998 occur again?

We don't know. Despite the fact that there has been cooling since 1998, you may be challenged by an activist teacher who would claim that warming continues, that there has been no recent cooling and that the 0.5°C is just the tip of the iceberg before catastrophic warming.

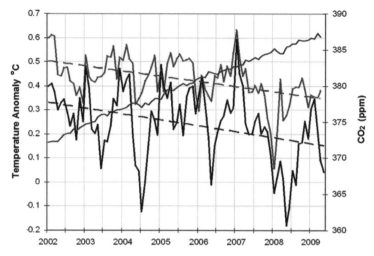

Figure 18: *Plot of 21st century temperature decrease (thermometer [Hadley CRUT3v, light grey] and satellite [UAH MSU, black]) against carbon dioxide increase showing a lack of correlation between temperature and carbon dioxide.*

6. If we have dangerous warming and the global temperature has increased by 0.8°C since the Little Ice Age, does this mean that the ideal temperature for life on Earth is that of the Little Ice Age?

During the Little Ice Age, people died like flies and it was really not a good time to be on Earth. Besides the cold, there were crop failures, famine, cannibalism and disease. As a child, you might have been on the menu. It was certainly not an ideal temperature then. However, a clever teacher would put you in your place and may suggest that the ideal temperature for an Eskimo is not the ideal temperature for someone living in the jungles of Borneo. You could then come back and suggest that this shows that humans can adapt to a great range of temperature so why worry about a warmer world.

7. **The temperature increase between breakfast and lunch is far higher than the 0.8°C temperature rise over the last 150 years. Why is such a small change over 150 years dangerous yet larger changes each day are not?** What can you teacher say? Maybe your teacher might argue that a 0.8°C temperature rise is a sign of terrible things to come. Who knows? Email me your teacher's answer to ian.plimer@adelaide.edu.au.

8. **If global warming is human in origin, when will we feel it and when will it be dangerous?** Considering that there is global cooling in the 21st century, the answer should really show you what your teacher is about. If it is "We can feel it now", then the teacher is telling you lies. It is neither felt nor measured. If you are told that it will be dangerous in 50 or 100 years, then just roll your eyes and turn off. Humans have adapted to far more rapid changes in climate and history tells you that in warmer times you are wealthier, you eat better, you have a better standard of living and you live longer. If you really want to worry yourself about climate, then think what it would be like when we have the next inevitable glaciation. Most of the world's Northern Hemisphere population would either be covered by kilometres of ice or live in a desert.

9. **In the last 100 years, has there been global warming and global cooling?** There has been both warming and cooling. If the answer is "I don't know", your teacher is honest. If the answer is a definite no, then your teacher is an activist trying to fill your head with nonsense.

10. **In the last 100 years, we have had global warming alternating with global cooling on 60-year cycles. Which part of the global warming in the last 100 years has been driven by human actions and which part is natural?** We don't know. However, because the rate of change of past warmings is the same as the

most recent warming, it is most likely that any human-induced global warming signal is swamped by natural changes. An honest teacher would say "I don't know". An environmental activist would claim that there has been no cooling, it has only warmed and all of this warming is of human origin. Any teacher that does not say "I don't know" is probably not good for your education.

11. Have past climate changes been greater and quicker than modern changes? Yes. Some massive changes have taken a few decades and changes have been ten times as great as the scariest predictions for our future. You should be quite happy if your teacher says "I don't know" and you should just roll your eyes with disbelief if your teacher says no. That is the unscientific view of an environmental activist.

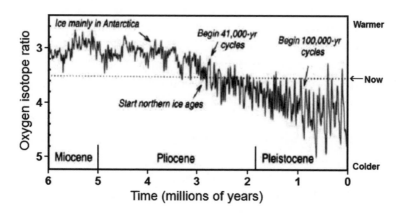

Figure 19: *Climate change over the last 6 million years using oxygen isotope proxies for temperature (Ocean Drilling Program site 677, North Atlantic Ocean). In the past, it has been warmer and temperature variation has been greater than at present. Climate cycles of 41,000 years were replaced by 100,000-year climate cycles 1 million years ago during a long cooling period.*

12. **Does the Sun drive warmings and coolings of the Earth?** It should come as no surprise that that great ball of heat in the sky has the major influence on global temperature. This is a test to see if your teacher is up to date and knows a bit about history. The answer is yes. This we see from the Medieval Warming when the Sun was active and from the very cold times in the Little Ice Age when a lack of sunspot activity coincided with very cold times. At present, there is a very interesting debate in science about exactly what the Sun is doing, whether we are heading for another very cold period and whether most of the last 150 years of warming is due to the Sun. Time will tell. If the answer from your teacher is no, then you should complain to the head teacher that your teacher is a buffoon. A follow-up question is in order. Try another question on climate and the Sun.

13. **Why is there no correlation between global warming and atmospheric carbon dioxide yet there is a correlation between solar activity and temperature?** "I don't know" at least would be an honest answer. There is no correlation because carbon dioxide does not drive surface temperatures, it is the Sun. Surprise, surprise. Your activist teacher can only dispute the data and argue that zillions of scientists have shown that carbon dioxide drives temperature, that there is a consensus and that solar activity may have a very small effect on climate. If your teacher does not stress the Sun, then your time is being wasted.

14. **Why do Mars and other planets show global warming?** It is because of the Sun. If you get a really silly answer from your teacher, then follow up with the next question which is silly.

15. **If it is not the Sun driving global warming on Mars, what industries on Mars are pumping carbon dioxide into the Martian atmosphere?** This question will get you smacked around the head, turfed out of class or expelled because you have made a total

fool out of your teacher in front of all your fellow pupils. If you didn't get a chance to ask this question then try another.

16. **Where does the carbon dioxide in the atmospheres of the moon and Mars come from?** Clearly it is not from human emissions of carbon dioxide. It is because rocky planets and their moons still leak out carbon dioxide, as the Earth does. Your teacher might want to say that these atmospheres are very thin, which is true, but this does not answer the question.

17. **Billions of years ago, did the Earth's atmosphere contain carbon dioxide?** The only answers are yes or "I don't know". Answers such as we could not possibly work this out are deceptive and shows that your teacher knows very little about Earth history. If the answer is yes, ask the next question.

18. **Where did the carbon dioxide in the Earth's atmosphere billions of years ago come from?** Volcanoes. This is the normal degassing process of this and other planets. Traces of carbon dioxide may have been added by comets. If your teacher claims that it was life that emitted carbon dioxide, then this is wrong. There were no animals on early Earth exhaling carbon dioxide.

19. **If there was a lot of carbon dioxide in the atmosphere of early Earth, where did it go?** It went into limestone, carbon-rich sediments and life. It certainly did not stay in the atmosphere otherwise the atmosphere would contain monstrous amounts of carbon dioxide. You might continue to put pressure on your teacher with a few more questions.

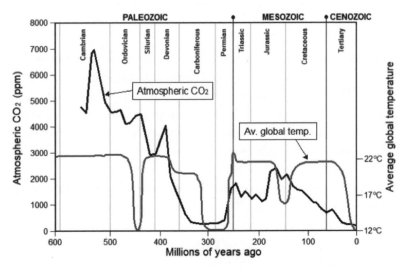

Figure 20: *Average global temperature plotted against atmospheric carbon dioxide (CO_2) over the last 600 million years. The four periods of low temperature were ice ages, global temperature in the current ice age is lower than the long-term past global temperature average and there is no long term relationship between atmospheric carbon dioxide and temperature.*

20. How much carbon dioxide does limestone contain?
Limestone contains 44% carbon dioxide because it is $CaCO_3$, the atomic weights are Ca=40, C=12 and O=16 and CaO is therefore 56 and CO_2 44. Now for the killer question.

21. Where did the carbon dioxide for limestone come from?
Originally from the air and then it was processed through the oceans and life to form limestone. Keep going, you have your teacher on the rack if you are allowed to continue to ask more questions. This might not happen as you are getting pretty close to showing that the teacher knows little about the present and past environments.

22. Does this mean that the air billions of years ago had more carbon dioxide than now? Yes. Before about 2500 million years ago, there was no limestone. Calcium magnesium carbonate (dolomite) rocks were common after that time and experiments show that dolomite can only form when there is a large amount of carbon dioxide in the air. With a lower atmospheric carbon dioxide content calcium carbonate forms and with an even lower atmospheric carbon dioxide content, calcium sulphate forms. If the answer given is no, the teacher is clearly aware of where the questions are going, wants to shut down questioning and avoid having their ignorance displayed.

23. If the planet originally had far more carbon dioxide in the air than now, why isn't the planet permanently very hot? The only answer is that carbon dioxide is not the major driving force for keeping the planet warm. Climate is very complex and to reduce warming to increases in the trace gas carbon dioxide reduces the whole climate story to absurdity. You can accept "I don't know" is an answer but any other feeble explanation shows that your teacher is an activist with no basic knowledge. Keep going on the same theme.

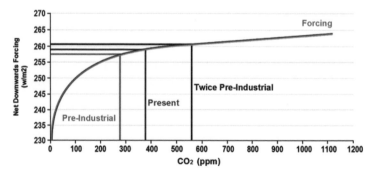

Figure 21: *The relationship between atmospheric carbon dioxide and forcing showing that with increasing atmospheric carbon dioxide, the warming effect decreases. The forcing in effect is temperature. The relationship between temperature and carbon dioxide shows that with increasing atmospheric carbon dioxide, the warming effect decreases. If the current atmospheric carbon dioxide content were doubled then there would be little effect on temperature. This is commensurate with former times when the atmosphere had hundreds of time more carbon dioxide than now yet there was no runaway greenhouse.*

24. If carbon dioxide drives global warming, how is it that we have had six major ice ages in the past yet atmospheric carbon dioxide was far higher then than now? This should make you extremely unpopular with an activist teacher. The aim of this question is to very quickly demonstrate that your teacher is an environmental activist using classes for political advocacy. How dare you asked a logical question based on knowledge that has been validated? Few school teachers have any knowledge of geology so the only way for an activist teacher to handle this question is to question your facts, slam you down, ignore you or throw you out for being disruptive. Don't think that you will get an answer to this question. The teacher might be silly enough to try to argue that it is just geology and that processes that happened million of years ago are too slow or do not operate today. Codswallop. The processes that operated in the past still operate today.

25. Will increased atmospheric carbon dioxide increase food production? Yes. This we know from measurements and experiments in glass houses. This is a great opportunity for your teacher to ramble on about limits to growth, depletion of resources and how we will all be doomed but none of this answers your question. An activist teacher would in no way want to admit that more carbon dioxide in the air would be good for people, especially those in the Third World. This is why many people claim that rabid environmentalism is anti-human and keeps the Third World in grinding poverty.

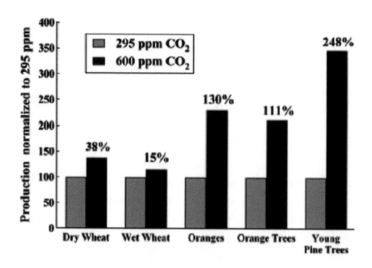

Figure 22: *Crop and tree yield increases resulting from natural fertilisation from an increase in atmospheric carbon dioxide. It is little wonder that horticulturalists pump warm carbon dioxide-rich air into glasshouses to increase crop yield.*

26. What drives climate change? This is a very complicated question. The best answer is that climate change derives from a combination of orbital, solar, terrestrial and extraterrestrial processes. Many times they interact. There are still many things we do not know about climate and new processes are continually being discovered and therefore the science is not settled. If the answer given by your teacher is carbon dioxide, then you know that you are exposed to political activism dressed up as education.

27. Can humans change climate? The natural forces are far greater than anything that humans can ever muster. We must remember that water vapour is the most powerful greenhouse gas and only one in 88,000 molecules in air is carbon dioxide of human origin. And then it only stays there for a short time before it is naturally sequestered. The only answer is no. Your activist teacher will say yes or probably.

28. How much carbon dioxide is in the atmosphere at present?
Such a simple question that few can answer. It is 389 parts per million by volume [0.039%]. Most people do not know that it is only a trace gas in the atmosphere and your teacher is probably one of these. See what your teacher can pluck out of the air. You should know that we humans breathe out 4% carbon dioxide so it clearly could not be toxic. If your teacher gets it wrong, stay quiet and think of how your time is being wasted. If the teacher gets it even half right or right, then go for the next question.

29. What proportion of annual carbon dioxide emissions derives from humans and what proportion is natural? This is a stupid question, unless you want to argue that humans are not natural. The IPCC states that only 3% of annual carbon dioxide emissions are from human activities which include emissions from industry, vehicles, coal-fired power stations, bushfires and changing land use patterns.

30. Can you please show me how the 3% of annual emissions of carbon dioxide, that is the human emissions, drive climate change and the other 97% do not? That's it. You will now be kicked out of class. The answer is, of course, that this has never been shown even though it underpins the story of human-induced climate change. If it has never been shown then the whole story of human-induced global warming is just nonsense. If your teacher questions this figure, then point out it is the IPCC's own figure. Your activist teacher will bluster on about consensus, the science is settled and generally resort to the authority of scientists and scientific organisations. The teacher may not understand that only one piece of evidence is needed to show that all the dogma is incorrect and then the hypothesis must be rejected.

31. Why don't the variations in atmospheric carbon dioxide correlate with human emissions of carbon dioxide? This one will put the cat amongst the pigeons. Human emissions clearly have very little effect on total carbon dioxide emissions showing that we really can't be dogmatic about the natural sources of carbon dioxide. This means that human emissions of carbon dioxide not only don't control climate but they don't even control global carbon dioxide levels. All your activist teacher can do here is to argue about the source and accuracy of the data, something that you are doing anyway with your questioning.

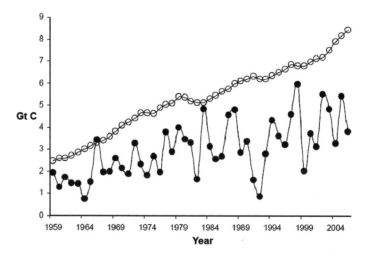

Figure 23: *Sources and sinks of carbon dioxide showing that human emissions of carbon dioxide (circles) are increasing at a greater rate than the increase in carbon dioxide in the atmosphere (solid circles). This suggests that there is either an unknown sink for carbon dioxide or that the oceans remove more carbon dioxide from the atmosphere than models indicate.*

32. Do plants know the difference between carbon dioxide emitted from human activities and carbon dioxide from natural emissions? Of course not. But, if your teacher starts to rabbit on about carbon isotopes then you know that this is an attempt to blind you with science and to avoid answering the question. This is a characteristic of the climate industry.

33. Is carbon dioxide poisonous? No. You may be told to try to breathe air with 20% carbon dioxide and see whether you think carbon dioxide is poisonous. This does not answer the question and treats you with disdain. If you try to breathe an atmosphere of 20% carbon dioxide or 95% nitrogen, you will suffocate. This is from a lack of oxygen and not from a poison.

34. Is carbon dioxide a pollutant or is it used in photosynthesis? A deliberately poorly constructed question. The answers are no and yes. The question has the ability to tie a teacher in knots because photosynthesis is one of the first things taught in school science. It shows that carbon dioxide is plant food and not a pollutant. Try the question. Your teacher may get it right, no teacher should say "I don't know" because it is elementary science and an environmental activist teacher may spend time trying to talk about too much or too little carbon dioxide, human emissions and pollution. Just listen and make up your mind.

35. Are atmospheric carbon dioxide levels a consequence of temperature, not the cause? This will rattle your teacher because the evidence certainly suggests that as temperature rises, the level of carbon dioxide also rises but a little later. Your teacher will probably wriggle and flatly deny this. However, the ice core data shows a different story.

36. If the human body and food are composed of carbon compounds and all animals breathe out carbon dioxide, how can carbon be pollution? The only honest answer to this question is that the term "Carbon Pollution" is deliberately used to mislead and deceive you. Any other answer is treating you like a fool. Carbon is black.

37. **For thousands of years, prophets of doom have been telling us the world was about to end. It hasn't, otherwise we would not be here. Why is it that we should believe the modern prophets of doom who tell us that our carbon dioxide emissions will destroy the planet?** If your teacher tries to tell you that this time it is different, then you know who is teaching you. Previous predictions were also based on mathematics, science, authority and consensus and previous predictions were used to extract money from credulous people. If your teacher tries to tell you that in the past there have been no predictions about the end of the world, then your teacher is far worse than you originally thought. Pack your bags and walk out.

38. **If we double the amount of carbon dioxide in the air from human emissions, how much will temperature increase?** This will really show whether your teacher understands anything about the role of carbon dioxide and global warming. If the teacher plucks a figure out of the air such as 2°C or 5°C, then you are being fed unsubstantiated environmental propaganda. If you want to make something up and sound convincing, then state it firmly, raise your voice a little and use the power of authority. You could politely ask for a reference that substantiates such a figure. The teacher may answer that this is the consensus, that this is the figure of the IPCC or that this has been shown by computer models. The relationship between carbon dioxide and temperature is logarithmic and the more carbon dioxide in the air, the less effect it has as the figure below shows.

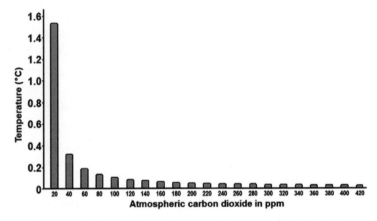

Figure 24: *The relationship between temperature and carbon dioxide showing that with increasing atmospheric carbon dioxide, the warming effect decreases. If the current atmospheric carbon dioxide content were to double, then there would be little warming. This is commensurate with former times when the atmosphere had hundreds of times more carbon dioxide than now yet there was no runaway greenhouse.*

39. Over the history of time, why has atmospheric carbon dioxide decreased? It has been naturally sequestered into sediments. This is a slight variation on the next question. This may be challenged but such a challenge only advertises the teacher's ignorance about geology.

40. Where has it all gone? Sediments. These have later been hardened into limestone, carbonaceous shale and other organic- and carbonate-rich sedimentary rocks. Life has also sequestered carbon dioxide and it is now stored in carbonate shell fossils, fossilised coral and algal reefs, coals and petroleum. There is far more carbon dioxide locked up in these rocks than in all the oceans, vegetation, life and the atmosphere put together. The atmosphere actually now contains a very small amount of carbon dioxide.

41. China's carbon dioxide emissions each year rise by five times the proposed cuts to Australia's emissions. Will Australia cutting carbon dioxide emissions change global climate? Of course not but this will open the floodgates for your environmental advocate teacher to talk about leading the way, being a good global citizen, population, morality and Australia being the example to change the ways of other countries. Don't even bother to ask the common sense question of why would China slow down its economic growth just because Australia has chosen to do so. In order of annual emissions of carbon dioxide we have China, USA, India, Russia, Japan, Germany, Canada, South Korea, Iran, UK, Saudi Arabia, South Africa, Mexico, Brazil and Australia. If Australia reduces emissions, do you really think that any of these 14 countries will also reduce emissions because of our great and unselfish sacrifice. If the world is not joining in, why should Australia bother? If the UK were to totally stop emissions, the global temperature would decrease in 2050 by $0.0239°C$. If the European Union totally stopped emissions the effect would be to reduce global temperature by $0.1767°C$. If the US totally stopped emissions, the reduction would be $0.2494°C$. A hell of a lot of pain for no gain.

42. Is water vapour or carbon dioxide the main greenhouse gas in the atmosphere? One answer might state that carbon dioxide is the main greenhouse gas. This is wrong. The main greenhouse gas is water vapour and it contributes to at least 95% of the greenhouse effect. Carbon dioxide contributes to about 4% and other gases about 1%. If you really want to get into trouble, then ask the next question.

43. If water vapour is the main greenhouse gas, why doesn't the government have a tax for water vapour emissions? The teacher will accuse you of being too smart for your own good may throw you out or try to bumble along trying to claim that carbon dioxide is somehow worse for the world than water vapour. Sit back, watch the fun and see how your teacher gets tied in knots. While on the water vapour theme, you might ask an elementary question that showed a prominent warmist to be a total dope.

44. **Does warming cause drought or does drought cause warming?** When warm weather evaporates water from moist soil, it requires energy to do this. The energy comes from the air and soil which then cool down. In drought when there is no soil moisture, the air is not cooled by evaporating water so it heats up. These are local and not global effects. In my opinion it was David Karoly, a prominent warming catastrophist and master of consistently wrong predictions, who showed that he does not understand elementary school science by claiming in a WWF report in 2003 that warming causes drought: "…the higher temperatures caused a marked increase in evaporation rates, which sped up the loss of soil moisture and the drying of vegetation and watercourses." Karoly was slammed in the scientific literature and still staggers around trying to tell us that we will suffer catastrophic global warming from our carbon dioxide emissions. Maybe if you attend one of his doom-and-gloom meetings or hear him on the radio, you could ask this question. I know what will happen, he will talk over you and talk you down but he will certainly not answer the question.

45. **How can you explain why it was warmer in the Medieval Warm Period than now yet there were no carbon dioxide-emitting industries?** This is a key question. You will probably get thrown out of class for this one. Your teacher might claim that there was no Medieval Warming [as Michael Mann tried to do with his fraudulent 'hockey stick']. This is dishonest. Your teacher might claim that the Medieval Warming was not nearly as intense as the data suggests. In fact, not only was it warmer than now but the Medieval Warming was global. This shows that the teacher does not know much. An even more bizarre claim might be that

we are in a period of cooling but human emissions of carbon dioxide have made it warm. There is no evidence for this and it does not explain both warmings and coolings in the 20th and 21st centuries. The explanation is simple: The Medieval Warming was natural and it is probable that with any warming today, the natural warming swamps anything that may be of human origin.

46. If the warming in Medieval times was natural, what is the evidence to show that late 20th century warming was not natural? This is a deliberately poorly constructed question because you can't prove a negative. Give it a go. If your teacher slams you for asking to prove a negative, then your teacher is using logic and is probably a good teacher. If your teacher tries to answer the question, then the teacher advertises that they can't think clearly and can't put together a decent argument. The answer given may be that humans are now emitting far more carbon dioxide into the atmosphere than in Medieval times. This is correct. However, this does not show that the 1860 to 1880, 1910 to 1940 and 1977 to 1998 warmings were of human origin and does not explain the 1880 to 1910, 1940 to 1977 and 1998 to the present coolings. More than likely a teacher will state that most scientists believe that humans produce global warming, that there is a consensus in the scientific world and those scientists with a contrary view are nutters paid by big oil or are discredited. If you get such an answer, then you really know that your teacher is a fool feeding you with propaganda and you should come in for the kill with the next question.

47. Since thermometer measurements were made, there has been warming from 1860 to 1880, cooling from 1880 to 1910, warming from 1910 to 1940, cooling from 1940 to 1977, warming from 1977 to 1998 and cooling from 1998 until now. Which warmings and coolings were of human origin? This is guaranteed to get you thrown out of class and possibly expelled. A dishonest teacher will dogmatically state that there has only been warming. An honest teacher will state that it is not possible to work out which warming was natural and which was of human origin. Alternatively, your teacher may correctly argue that since records were kept there has been indeed warming and cooling but the total warming has been in the order of 0.8°C.

48. Why could the Northwest Passage be navigated in the 1930s and 1940s in wooden boats yet it could not be navigated in the late 20ᵗʰ century warming? Again, you will make the teacher very uncomfortable. There are very good records that the Royal Canadian Mounted Police used to patrol the Northwest Passage.

49. In 1903 Amundsen passed through Canada's Northwest Passage from the Atlantic Ocean to the Pacific. If the planet is warming, why is this not possible now? Again, a somewhat unfair question but who cares. The teacher should know that the Arctic sea ice extent changes on 30 and 18.6 year cycles and that this has nothing to do with alleged human-induced global warming.

50. I heard that 2010 was the hottest year since records have been kept. The Northwest Passage is closed by ice yet it was open in the 1930s. Was 2010 really the hottest year on record? No. All your teacher can do to answer this question is to quote news articles uncritically. This is a good test to see if your teacher believes what they read and hear or whether your teacher critically evaluates news that agrees with their political position. Just accept the answer in silence.

51. Why has the temperature been decreasing since 1998 yet human emissions of carbon dioxide have been increasing? The only thing your activist teacher can do is to question whether the El Niño of 1998 has created a bias in the results or to paraphrase a quote of Phil Jones, a Climategate luminary, who stated that there has been no statistically measurable warming since 1995. What the question asks is that although there has been a huge increase in human emissions of carbon dioxide over the last 15 years, there has been no corresponding increase in temperature. Ask the question and sit back and watch the fun. You might be correctly told that there are other great forces that drive climate such as the Sun and these dominate over human emissions of carbon dioxide. Look back at Figures 14 and 18, they give the story.

52. **What is the mechanism that causes warming trends to reverse? Will that mechanism kick in with human-induced global warming?** The Sun. Orbital oscillations place the Earth closer or further from the Sun. The Sun affects the amount of cosmic rays that reach the Earth's surface and can create more low-level clouds. The more low level clouds there are, the cooler it is. Accept the answer "I don't know". A really stupid answer might be that the atmospheric carbon dioxide content decreases during cooling and, with an increase, drives warming. If your teacher is an unreconstructed warmist, then the question cannot be answered and you have shown that your teacher is feeding you propaganda.

53. **The Earth has been warming since the Maunder Minimum 330 years ago. Is it surprising that temperature measurements would show a warming trend over the last 150 years?** If the answer is a definite no, then you are being well taught. If there is an "if", "but", "maybe" or your teacher argues about the question, then your teacher is ignoring data and feeding you propaganda.

54. **How can there be a global average temperature?** There can be if it is defined. However, it really means nothing as it is dependent upon measurement. This is a question to test whether your teacher knows about the limitations of measurement and whether your teacher knows the process of putting together temperatures around the world. Bear in mind that all temperature compilations are "adjusted".

55. What is the order of accuracy of temperature measurements? The first thing you should have learned in school science is order of accuracy. If you measure a temperature of 20°C, it is not exactly 20°C. With a modern thermometer it is 20.0 ± 0.1°C which means that the temperature is somewhere between 19.9°C and 21.1° C. All measurements have an order of accuracy or uncertainty. With historical measurements, the order of accuracy would have been about 1.3°C which means that the 20°C temperature measurement was somewhere between 18.7°C and 21.3°C. If temperature since measurements were first taken shows an increase of 0.8°C, then until there were far better thermometers the uncertainty was greater than the suggested temperature rise. This means that there was no detectable temperature rise. A teacher with a physics background would immediately tell you that the order of accuracy was half that of the smallest graduation on the thermometer [i.e. it is 0.5 to 0.1°C]. An environmental advocate may say that these are things that you should not worry about or may resort to authority and state that temperatures have been measured for a long time, these measurements are accurate and we check them many times.

56. Is it valid to combine inaccurate 19th century temperature measurements with far more accurate measurements of the late 20th century? No. But it is done. Your environmental activist teacher might prattle on about how older measurements have been checked, how there are mathematical ways of checking the accuracy and how there are oodles of scientists who accept these measurements. Furthermore, 90,000 19th and early 20th century carbon dioxide measurements are rejected by the climate industry, possibly because they show huge variability well above the order of accuracy of measurement. Just sit back and yawn. You do it in every other class.

57. The number of measuring stations has greatly decreased over the last 20 years with the loss of stations in polar, mountainous, rural and remote areas. Does this create a warming bias to temperature measurements? The correct answer is yes but we don't know how much. Your teacher might talk about better technology, remote temperature measurements, correction factors and consensus of scientists to avoid admitting that with fewer measuring station, the results are less reliable. You can check out how many stations have been lost from the University of Delaware animation of station locations over time [http://climate.geog.udel.edu/~climate/html_pages/air_ts2.html]. GISS also shows the lost stations [http://data.giss.nasa.gov/gistemp/station_data/].

58. When temperatures are used for models, are the actual raw measurements used or are corrected measurements used? This will make you really popular. If the teacher does not know, then fair enough. If the teacher knows that all measurements are "adjusted" then the teacher will know that the ground is very weak. Except, of course, if your teacher is not with it. If your teacher waffles on about corrections or the urban heat island, then come back with another question.

59. What is the urban heat island effect? Sit back and wait. The teacher might be honest enough to say "I don't know". Your teacher should know that temperature measurements in urban areas should be adjusted downwards because of the heat that cities produce. Your environmental activist will either not know or will talk about corrections, international standards, protocols, consensus and so on…you know the story. Just turn off and count the fly marks on the ceiling.

60. Is there a standard method to correct for the urban heat island effect? Your activist teacher will again talk about protocols, consensus and all sorts of things to avoid answering the question truthfully. There is an equation by Oke [Urban heat-island warming= 0.317lnP, where P=population] for corrections of temperature measurements in populated areas and there are scientific arguments about whether this equation tells the real story. It should be easy to make up your mind whether the teacher knows anything at all about human-induced global warming.

61. If land temperature measurements have to be "adjusted", how do we know that the "adjusted" measurements are accurate? This should raise the blood pressure of environmental activist teachers who can only resort to arguments like "trust me" or arguments of authority.

62. If "adjusted" temperature measurements are used for computer predictions of future climate, how can we trust these models? You may be told by activist teachers that these models are tested and tested and tested and shown to be correct. This is nonsense but listen politely before your next question.

63. Why have computer models that predicted high altitude warm air at the equator as a result of increased human emissions of carbon dioxide failed? This will get you into real trouble. Look at the diagram with the model and the observed balloon measurements that test the model. The model failed. Miserably.

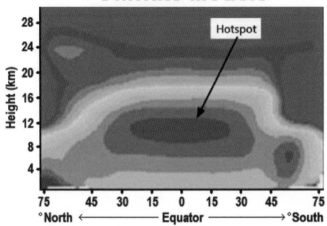

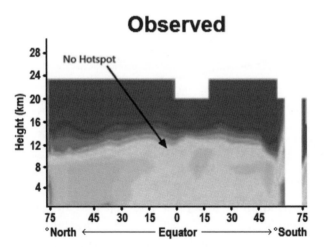

Figure 25: *Upper diagram shows predicted hot spot in the atmosphere at the equator at 10-12 km altitude from computer modelling of the effect of increased atmospheric carbon dioxide on the atmosphere. The lower diagram shows the measured atmospheric temperature from radiosonde balloons. The models of the atmosphere clearly do not correlate with the atmospheric observations.*

211

64. Computer models predicted that the sea surface temperature would warm yet measurements show that there is cooling. Why? You realise of course that computer models are the only basis for the idea that human emissions of carbon dioxide drive global warming but does your teacher know this? Does your teacher really know that the models are hopelessly wrong and the measurements tell the story? If your teacher waffles on about how good the models are, that mathematical models cannot be wrong, that there is a consensus and that so many scientists cannot be wrong, maybe you could ask a few more questions.

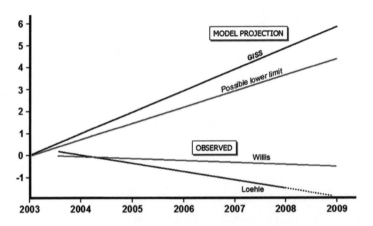

Figure 26: *Modelled sea surface temperature (°C change) from GISS compared to measured sea surface temperature changes from Willis Island (Coral Sea) and compilations by Loehle (2009). Models of sea surface temperature do not correlate with sea surface temperature measurements.*

65. What information is accepted and rejected in models of future climate? One thing for sure, your teachers do not know how the models work. Maybe the modellers don't either. No teacher would know what is considered or not considered in a model. If models are the only way in which the effects of increased carbon dioxide on climate are evaluated, then we need to have far more confidence that these models actually work.

66. **Computer models predict future climate changes far less than the changes experienced by humans over the last 6,000 years. Why should we worry about future climate?** This one is guaranteed to flush out the environmental activist. Expect answers about responsibilities, morality, duty and living in harmony with nature and don't at all expect the question to be answered.

67. **The models, code and data used by the IPCC for their climate predictions are not available yet computer climate predictions are the basis for suggestions of human-induced global warming. How can their predictions be independently checked?** This gets to the nub of the matter. Your environmental activist teacher can only answer that this is not true [Check the 2007 IPCC AR4 for yourself. It is true]. Maybe your teacher might claim that the computer predictions were done by eminent reliable scientists and hence they don't need checking. A quick additional question asking the teacher to name these people would put that answer to bed.

68. **How can I have confidence in the predictions of a climate model if I don't know how it works and if they have been shown to be wrong?** See what your environmental activist teacher does with this with question. Maybe they will say "Trust me" or "Trust the models of wonderfully eminent people". Maybe your teacher might again state that there is a consensus and so many scientists could not be wrong. You could ask: Who are these scientists? I doubt if Tim Flannery for instance has a clue about the limitations of climate models.

69. **Would you expect a warm climate after the Little Ice Age?** Yes. Is anyone really surprised that we have had over 330 years of natural warming after the Little Ice Age ended 330 years ago. Or would you expect cooling after a Little Ice Age. Get real.

70. **Since the depth of the Little Ice Age 330 years ago, the Earth has been warming. Which part of this warming is natural?** Your teacher may dispute whether there was actually a Little Ice Age. This shows ignorance and you should think about changing schools. Your teacher might be silly enough to try to argue that the planet had a wonderfully even climate until the Industrial Revolution 150 years ago and then we humans started to pump out carbon dioxide emissions and change the climate. Both answers show that your teacher is an activist with no knowledge of history. You should just pack your bags and walk out because you would learn more hanging around a shopping centre than listening to this teacher. Your teacher might beat you to it and throw you out. The answer is that it is only the warming period from 1977 to 1998 that is claimed to be of human origin and this is in dispute. If this is the answer given, then ask the next question.

71. **If most of the last 330 years of warming is natural, why isn't all of the latest warming natural?** You are now well down the path to sending your teacher nuts. The only honest answer is "It probably is but I really don't know". Time to change tack before the teacher has to take stress leave.

72. **Why are there 60-year cycles of warming and cooling over the last 2,000 years?** An activist teacher might argue about the data. This shows ignorance and you should let your mind wander into fantasyland when this teacher speaks because you will learn nothing. An honest teacher would say "I don't know". The answer is that there are shifts in the ocean every 25 to 30 years, the oceans carry far more heat than the atmosphere and so shifts in the ocean heat balance change air temperature.

73. **During ice ages, do we get cycles of warm interglacials and cold glaciations?** Yes. This is the only answer. If there is waffle about global warming, human emissions of carbon dioxide or that ice ages are in the past cannot be used as a guide to present times,

then your teacher is an ignorant activist fool. Turn your brain off and start to think about the weekend. If the answer is yes, then ask the next series of questions.

74. Why do we get cycles of glaciation and interglacials? Simple question. The simple answer is that the Earth's orbit is not perfectly circular and sometimes we are closer to the Sun and other times we are further away. Orbit cycles are such that at present we have about 90,000 years of glaciation and 10,000 years of interglacials. There are other orbital cycles. During glaciation, temperature fluctuates wildly whereas in interglacials, temperature fluctuates only slightly. A teacher might say "I don't know" and, even though they should know, you can be comforted by the fact that they are honest. An activist teacher feeding you propaganda will bluster and bully you and make comments about how times today are different because we humans are emitting carbon dioxide into the atmosphere. The world is already full of fools and this would just show that your teacher is one of them.

75. Would you expect a warm climate in an interglacial? Well of course we would. The current interglacial has been shorter and cooler than previous interglacials in the current ice age. This interglacial is nothing special. Just wait to see what your teacher says. Your teacher might state that human activities will extend the current interglacial, that human activities may push us out of the interglacial-glacial cycle or human emissions may produce irreversible catastrophic global warming.

76. Ice cores show a saw-tooth interglacial-glacial pattern with huge temperature variations in cold times. Why does temperature reach a maximum then fall and why were all past temperature maxima about the same? Sounds like a hard question but it is not. The saw tooth pattern derives from orbit cycles of 90,000 years of cold [glaciation] and 10,000 years of warmth [interglacial]. The reason why the temperature does not run

215

away to hotter and hotter conditions is because both evaporation and precipitation of water involve the transfer of energy and this operates as a temperature buffer. This is why in the past when carbon dioxide has been far higher than now, we did not have runaway global warming.

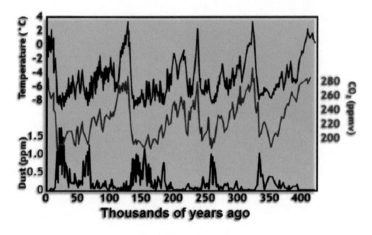

Figure 27: *Reconstruction of temperature from Vostok (Antarctica) ice cores showing interglacials and glaciations on a coarse scale where atmospheric carbon dioxide mirrors the temperature trend. On a finer scale, carbon dioxide peaks hundreds to thousands of years after temperature peaks. Climate changed on 100,000-year cycles with 90,000 years of glaciation and 10,000 years of interglacials. The increased amount of dust trapped in the ice sheets indicates that in cold times there was devegetation, strong winds and desertification. It is counter intuitive that deserts form in cold times. This we see throughout the history of the planet.*

77. **Ice core records show that carbon dioxide in the air increases 800 to 2,000 years after a natural event of global warming. Does temperature drive an increase in carbon dioxide or do the ice cores show that an increase in carbon dioxide drives temperature?** Again, you have given your teacher no way out. The evidence very clearly shows that an increase in temperature leads to an increase in carbon dioxide. The only way an environmental activist teacher can handle this question is to waffle on about

consensus, that the science is settled, that many great prestigious academies support the current popular view and that times are different now because we humans are emitting carbon dioxide into the atmosphere. Don't ever try to change the mind of someone by using logic when that person has come to a conclusion by not using logic. Sit back, listen to the nonsense and make up your own mind about your teacher.

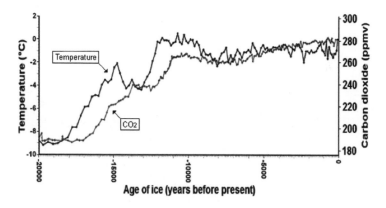

Figure 28: *Ice core data plot from Antarctica (Epica Dome C) of temperature versus carbon dioxide (CO₂) showing that rises in temperature stopped hundreds to thousands of years before carbon dioxide rises started. This plot shows that for the last 18,000 years, carbon dioxide has not driven warming and the inverse occurs: Warming drives an increase in carbon dioxide.*

78. **Does sea level rise and fall?** The only answer is yes. If the teacher gives any other answer, then close your eyes and give up. In the current ice age, sea level has risen and fallen by about 130 metres between glaciation and the peak of the interglacial in the current ice age. In past ancient ice ages, the sea level rise and fall between glaciations and interglacials was up to 1500 metres.

79. **Does the land level rise and fall?** The only answer is yes. However, you may get some quibbles about how fast land rises or falls. From the Roman port of Ephesus [now 15 kilometres inland], we

know that it is quick. This is validated by the ancient city of Lydia, now covered by water. Northern Scandinavia has risen at least 340 metres in 12,000 years and this land level rise is nearly three times as great as the sea level rise that took place over the same time period. Land level rise can be very quick. If your teacher states that natural land level rises and falls are very slow and far slower than sea level changes, then you are being fed nonsense.

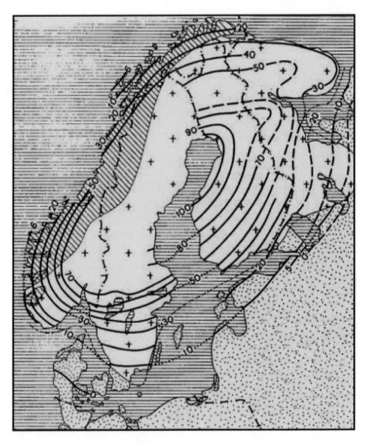

Figure 29: *Contours showing the amount of uplift (in metres) over the last 6,000 years in Scandinavia (diagonal lines, folded rocks of the Caledonides; crosses crystalline rocks; dots shallow marine sediments; horizontal lines are water). Scandinavia was pushed down by a 5-kilometre thick ice sheet, this has melted and the area is now rebounding. The maximum uplift rate is 60 centimetres per year.*

80. **If sea level goes up and down and land level goes up and down, how is global sea level measured?** The answer is: With great difficulty. There need to be a large number of tidal measuring stations all over the world that have been regularly and accurately surveyed and this information must be tied into land surveys. Tidal measuring stations can sink, which make the measurements look like a sea level rise. Satellites measure gravity, from which the theoretical shape of the Earth is calculated and from which in turn the theoretical sea level can be calculated. These calculations don't agree with tidal station measurements and, by changing a few assumptions in the satellite calculations, it can easily be shown that sea level is rising fast, is falling or is static and hence the desired model can be obtained. What we can measure is what sea level has done in the recent past.

81. **Why did the rate of sea level rise double as soon as satellites started to measure sea level?** Tide gauges show the current post glaciation sea level rise is 1.6 millimetres/year. As soon as satellites started measuring sea level from 1993, the sea level rise became 3.3 millimetres/year. Satellites measure gravity, not sea level, and a large number of calculations and adjustments must be made to deduce sea level changes from gravity measurements. Just one little adjustment to the computer and you can get whatever number you want.

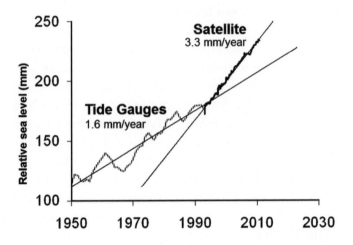

Figure 30: *Relative sea level with rise measured from tide gauges (1.6 millimetres per year) compared with satellite measurements of gravity computed to an interpreted sea level rise of 3.3. miilimetres per year. Sea level suddenly appeared to rise once the measuring technique was changed.*

82. Did sea level rise in the 400-year long Medieval Warming or the 600-year Roman Warming? Probably no. Just because temperature rises does not necessarily mean that sea level will rise. This you can flesh out in the next questions.

83. Is sea level rising now? Yes. By how much we don't really know. There are greatly variable and exaggerated figures that come from the climate industry and when people actually measure long-term sea level changes, then these are far lower than anything that computer models spit out. If I have a choice, I prefer to use measurements rather than numbers spat out from a computer. A computer answer depends on the information it is given to work with and if the information is limited, then the answer from the computer will be wrong.

84. Would we expect sea level to rise in an interglacial? Yes. However, the sea level rise is not well correlated with temperature changes so we have to be a little careful about arguing for a close correlation between sea level rise and temperature rise. God knows how your warmist environmental activist teacher would answer this question.

85. How much has sea level risen in the current interglacial and how long did this take? The latest sea level rise of some 130 metres started 12,000 years ago and, in most parts of the world, slowed down or completely stopped 6,000 years ago. In some parts of the world [e.g. the Great Barrier Reef or the Maldives], the land level is rising so it looks as if the sea level is falling, in other parts of the world the land level is sinking [e.g. south-eastern England] and on a global scale it looks as if sea level is rising very slowly.

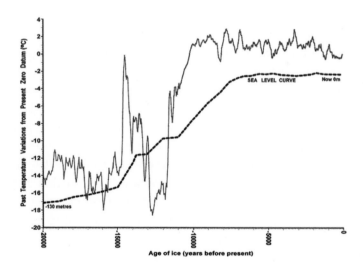

Figure 31: *Post-glacial sea level rise of 130 metres (black dashes) and temperature changes (sold grey line; derived from ice core measurements) showing (a) lack of good correlation between temperature and sea level change (b) temperature decrease since the Holocene Maximum 6,000 years ago (c) very rapid natural global warming after the cold Younger Dryas 11,300 years ago and (d) the slight variations in temperature and sea level in modern times compared to the past.*

86. **Why does sea level rise?** This is where you will find out if your environmental activist really knows anything about sea levels. If you teacher says from melting glaciers and expanding seawater in the warmer conditions, then this is only partly right. Because rocks are plastic and the ocean crust is thin, loading up the oceans with more water can actually make some parts of the ocean floor sink. In shallow water, the expansion of water has no effect on sea level. Sea level can rise when large masses of rock are pushed up on the ocean floor and these displace water. Sea level near mountains, in meteorological lows and gravity lows is higher than other places. Sea level can change with very slight changes in the rotation of the Earth and if delta, beach and continental shelf sediments compress, subside or lose fluids, then there can be a local sea level rise. Tidal cycles every 18.6 years create changes of sea level as does water slopping around from one side to the other in ocean basins.

87. **Is the land sinking at Venice or is Venice being flooded because of sea level rise?** Venice is sinking at about 7 centimetres per century. There are hundreds of years of tide records at Venice that show that there is a constant rate of sinking. If there was recent sea level rise due to human-induced global warming, then the tide gauges should show an increased rate of sinking [i.e. the combination of a constant rate of sinking plus sea level rise]. The records show no such thing. From the 1920s to the 1970s, Venice sank faster, probably as a result of groundwater extraction. If your teacher says this is a good example of sea level rise, then come back with another question and ask why all coastal cities in the world are not being flooded.

88. **Is the land rising in eastern Australia?** Yes. There has been a 2-metre land level rise over the last 5,000 years. What is your activist teacher going to say? Yes. No. I don't know. Or maybe claim that although there has been land rise, the sea level rise has

been faster? If this claim is made, you might ask about why there are raised beaches, raised rock platforms and raised oyster beds along the eastern Australian coast.

89. Why are there old beaches in the Murray-Darling Basin hundreds of kilometres from the modern shoreline and over 100 metres above sea level? One of the best examples of this is the Ginko mineral sands deposit at Pooncarie, NSW. An old beach is mined for the materials you use in tile glazes and as a non-toxic whitener in paint. These beaches, although millions of years old, show that there have been either great sea level changes or great land level changes. It is actually both. Because so few teachers know about geology, it will be interesting to get the answer. Maybe the facts will be challenged, demonstrating the teacher's ignorance, maybe the answer will be "I don't know" or maybe the answer will be that what happened in the past has nothing to do with the present. However, yesterday's present is today's geology.

90. If sea level has risen to separate Tasmania from Victoria, is the same natural process still in operation? Yes. Measurement of water depth shows that there are continuous stretches of the present Bass Strait that are less than 120 metres water depth and hence it would have been possible to walk to Tasmania before 12,000 years ago. But what on Earth will your environmental activist dream up to answer this question? You can only wait and see. I suspect that you will be beaten around the head with consensus, that you will be told that these processes might once have been dominant but now human-driven climate change rules the world and that all scientific academies support the view that humans are driving climate change and the resultant sea level rise. This is untrue and the Japanese, Russian, Indian and other academies have not supported this nonsense.

91. **How did aboriginal people get to Australia before boats were invented?** They walked from Irian Jaya during the last glaciation when sea level was 130 metres lower than now.

92. **Which part of the present sea level rise is due to post-glacial sea level rise and which part is due to human activity?** The answer is we don't know but the example from Venice shows that human-induced sea level rise is extremely small. It will be interesting to see how an environmental activist school teacher would try to claim that all the sea level rise is due to human activities because this would mean accepting that sea level rise from 12,000 to 6,000 years was natural and then, after the Industrial Revolution 150 years ago, the natural processes were suddenly switched off. Pull the other leg.

93. **Why do Al Gore and Tim Flannery tell us that sea level will rise more than 8 metres and yet they have expensive waterside properties?** This is a Friday afternoon question so you can get tossed out of class and get home early. I cannot see how your environmental activist school teacher can continue to support these doom-and-gloom catastrophists who show such hypocrisy. Your teacher's answer may challenge whether their gurus really do live at sea level or maybe the teacher maybe provide some lame excuse. Email me the answer.

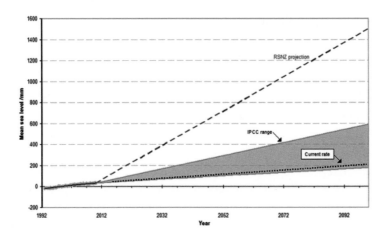

Figure 32: *Plot of mean sea level versus time showing the current and projected current rate of sea level rise (dots), the Royal Society of New Zealand projection (dashes) and the speculated IPCC range (shaded) for future sea level. There is no evidence to show that there was a sudden sea level rise in the 21st century and the RSNZ and IPCC projections are just speculations.*

94. What is pH? It stands for *pondus Hydrogenii,* is the measure of the negative log of the hydrogen ion activity in a solution. Neutral solutions have a pH of 7, solutions below a pH of 7 are acid, and those above a pH of 7 are alkaline. The important thing about pH is that it is a logarithmic scale.

95. What is the pH of the oceans? The oceans are alkaline. They pretty well always have been. Get used to it. The pH varies from 7.5 to 8.4. In some places near submarine hot springs the oceans are slightly less alkaline. Monstrous amounts of acid would have to be added to the oceans to lower pH in seawater by 0.1 units.

96. What is a buffer? A buffer adjusts the pH of a solution. Seawater has a large number of chemicals that buffer pH and reactions between seawater and minerals in submarine rocks and sediments such as feldspars also buffer seawater. Because seawater is buffered, additional atmospheric carbon dioxide will dissolve as carbon dioxide, bicarbonate and carbonate and buffers will keep seawater alkaline. It has always worked this way, even when there was far more carbon dioxide in the atmosphere and it is hard to see why basic chemistry would change just because you and I are alive.

97. **Why haven't the oceans become acid in past times when atmospheric carbon dioxide was high?** If oceans were acid, then calcium carbonate would have dissolved, we would have had periods of time when limestone was not deposited and we would have had periods of time when no calcium carbonate shells would have been fossilised. The global fossil record has no gaps where carbonates are totally absent. The seas must have been alkaline throughout time. We have not had acid oceans despite the fact that in the past atmospheric carbon dioxide was far higher than now. There are various other chemical fingerprints in seawater to show that it has been alkaline for the last few billion years. With more carbon dioxide in the air, more dissolves in ocean water. This carbon dioxide exists as dissolved gas, bicarbonate ions and carbonate ions. The proportion of these three forms of carbon dioxide changes with temperature, pressure and the chemistry of seawater. Your environmental activist teacher would need to know a bit about chemistry to give you the answer so expect to be baffled by waffle. The correct answer is that both seawater and seafloor sediments and rocks are buffers. Each year at least 10,000 cubic kilometres of seawater passes through ocean floor volcanic rocks. This seawater circulates to at least 5 kilometres beneath the seafloor and chemical reactions between seawater and the rocks keep seawater alkaline. These chemical reactions have been copied in laboratory experiments.

98. **Is there any activity in my life that does not involve the emission of carbon dioxide?** Everything you do or use in life ultimately involves emissions of carbon dioxide. If you want to avoid emitting carbon dioxide into the atmosphere then it is simple. Drop dead.

99. Why do those advocating human-induced global warming vilify scientists who disagree rather than addressing genuine scientific questions? The answer is of course that those in the climate industry do everything in their power to avoid facing questions, having to justify their position or debating someone who actually might know something about the subject. Your environmental activist teacher will try to argue that there is a consensus and the science is settled, will ask you why do very important people want to waste their time with nutters and will say that those few who disagree are not really scientists anyway or are employed by evil capitalists. Even at sober low profile scientific conferences such as the May 2011 GAC-MAC conference in Ottawa, there was a session on human-induced climate change. Warmists refused to attend which led to the participants wondering why warmists refused to address their scientific peers out of sight of the public and the media. The gathered scientists were left to conclude that the warmists probably have no testable scientific arguments to support their position.

100. What funds does the Australian government give for grants annually to show the existence of human-induced global warming? At least $142 million at last count but likely to be far more. However, when it is all added up, billions of dollars have been spent to find a problem. And the result? No problem has been found but the funding dollars keep rolling in.

101. What funds does the government grant to consider the possibility that there might not be human-induced global warming? None. Your environmental activist teacher will become apoplectic at such a question so now is your chance.

A guide for teachers

The mind is like a parachute. It only works when it is open. The greatest satisfaction a teacher can have is to open the mind of a young person, instill the joy of learning and knowledge, to teach young people to use critical and analytical thinking and to be a moral, intellectual and spiritual mentor to young receptive minds. Young people need to know that they should not believe what they are told, and what they read or hear without validation. To teach activism or the current political dogma does not prepare people for the world and is an abrogation of the calling of teaching. And what are the rewards of good teaching? Some 25 years after teaching someone, you will receive a moving email, letter or phone call. This person will say:

> *"I have always wanted to contact you and now I have finally done it. I just wanted to say that you said this and that and you changed my life. Thank you".*

When this happens there is always just a slight problem. You cannot remember who the person is or what you said so many years ago.

228

REFERENCES

This book is the sequel to *Heaven and Earth* in which there were thousands of references. *How to get expelled from school* is underpinned by the thousands of references cited in *Heaven and Earth*, presents material not cited in *Heaven and Earth* or published since *Heaven and Earth* appeared and used in *How to get expelled from school* and provides useful web links. These web links show that there is a huge body of science contrary to the popular paradigm that there are many scientists who give a different picture to that we hear in the media and there is no consensus in science on matters climate. Other sites give contain the raw information used for doom-and-gloom predictions.

Bell, R. E., Ferraccioli, F., Creyts, T. T., Braaten, D., Corr, H., Das, I., Damaske, D., Frearson, N., Jordan, T., Rose, K., Studinger, M. and Wolovick, M. 2011: Widespread persistent thickening of the East Antarctic ice sheet by freezing from the base. *Science* DOI:10.1126/science.1200109

Bird, B. W., Abbott, M. B., Vuille, M., Rodbell, D. T., Stansell, N. D. and Rosenmeier, M. F. 2011: A 2,300-year-long annually resolved record of the South American summer monsoon from the Peruvian Andes. *Proceedings of the National Academy of Sciences* 108: 8583-8588

Casenave, A. and Llovel, W. 2010: Contemporary sea level rise. *Annual Review of Marine Science* 2: 145-173

Church, J. A., White, N. J. and Hunter, J. R. 2006: Sea-level rise at tropical Pacific and Indian Ocean islands. *Global and Planetary Change* 53: 155-168

DelSole, T., Tippett, M. K. and Shukla, J. 2010: A significant component of unforced multidecadal variability in the recent acceleration of global warming. *Journal of Climate* 24: 909-926

Dragi, A., Anicin, I., Banjanac, R., Udovicic, V., Jokovic, D., Maletic, D. and Puzovic, J. 2011: Forbush decreases – clouds relation in the neutron monitor era. *Astrophysics and Space Science Transactions* 7: 315-318

Dressler, A. E. 2010: A determination of the cloud feedback from climate variations over the past decade. *Science* 330: 1523-1527

Funder, S., Goosse, H., Jepsen, H., Kaas, E., Kjaer, K. H., Korsgaard, N. J., Larsen, N. K., Linderson, H., Lysa, A., Möller, P., Olsen, J. and Willerslev , E. 2011: A 10.000-year record of Arctic Ocean sea-ice variability – View from the beach. *Science* 333: 747-750.

Gerlach, T. 2011: Volcanic versus anthropogenic carbon dioxide. *EOS* 92: 24.

Gratiot, N., Anthony, E. J., Gardel, A., Gaucherel, C., Proisy, C. and Wells, J. T. 2008: Significant contribution of the 18.6-year tidal cycle to regional coastal changes. *Nature Geoscience* 1: 169-172

Grinsted, A., Moore, J. C. and Jevrejava, S. 2009: Reconstructing sea level from paleo and projected temperatures 200 to 2100 AD. *Climate Dynamics* doi: 10.1007/s00382-008-0507-2

Hillier, J. K. and Watts, A. B. 2007: Global distribution of seamounts from ship-track bathymetry data. *Geophysical Research Letters* 34: doi: 10.1029/2007GL029874

Houston, J. R. and Dean, R. G. 2011: Sea-level acceleration based on U.S. tide gauges and extensions of previous global-gauge analyses. *Journal of Coastal Research* 27: 409-417

Idso, C. D. 2009: *Global warming and coral reefs.* Vales Lake Publishing, Colo, USA.

INQUA 2000: Sea level changes and coastal evolution. www.pog.su.se

Jaramillo, C., Ochoa, D., Contreras, L., Pagani, M., Carvajal-Ortiz, H., Pratt, L. M., Krishnan, S., Cardona, A., Romero, M., Quiroz, L., Rodriguez, G., Rueda, M. J., de la Parra, F., Morn, S., Green, W., Bayona, G., Montes, C., Quintero, O., Ramirez, R., Mora, G., Schouten, G., Bermudez, H., Navarrete, R., Parra, F., Alvarn, M., Osorno, J., Crowley, J. R., Valencia, V. and Vervoort, J. 2010: Effects of rapid global warming at the Paleocene-Eocene boundary on neotropical vegetation. *Science* 330: 957-961

Joughin, I., Smith, B. E. and Holland, D. M. 2010: Sensitivity of 21st Century sea level to ocean-induced thinning of Pine Island Glacier, Antarctica. *Geophysical Research Letters* 37: doi: 10.1029/2010GL044819

Katsman, C. A. and van Oldenborgh, G. J. 2011: Tracing the upper ocean's "missing heat". *Geophysical Research Letters* 38, L14610, doi:10.1029/2011GL048417

Konikow, L. F. 2011: Contribution of global groundwater depletion since 1900 to sea level rise. *Geophysical Research Letters* 38: doi:10.1029/2011GL048604

Kopp, R. E., Simons, F. J., Mitrovica, J. X., Maloof, A. C. and Oppenheimer, M. 2009: Probabilistic assessment of sea level during the last interglacial stage. *Nature* 462: 863-867

Lee, H. F. and Zhang, D. D., 2010: Changes in climate and secular population cycles in China, 1000 CE to 1911. *Climate Research* 42: 235-246

Lindzen, R. and Yong-Sang Choi, Y. 2011: On the observational determination of climate sensitivity and its implications. *Asia Pacific Journal of Atmospheric Sciences* 47: 377-390

Long, E. 2010: Contiguous U.S. temperature trends using NCDC raw and adjusted data for one-per-state rural and urban station sets. http://science-andpublicpolicy.org/images/stories/papers/originals/Rate_of_Temp_Change_Raw_and_Adjusted_NCDC_Data-pdf

McLean, J. 2010: We have been conned: An independent review of the Intergovernmental Panel on Climate Change (IPCC). Science and Public Policy Institute, August 18th 2010

Menne, M. J., Williams, C. N. and Palecki, M. A. 2101: On the reliability of the U.S. surface temperature record. *Journal of Geophysical Research* 115: doi:10.1029/2009JD013094

Miall, A. D. and Miall, C. E. 2009: The geoscience of climate and energy: Understanding the climate system and the consequences of climate change for the exploitation and management of natural resources: The view from Banff. *Geoscience Canada* 36: 33-41

Morner, N.-A. 2010a: Some problems in the reconstruction of mean sea level and its changes with time. *Quaternary International* 221: 3-8

Morner, N.-A. 2010b: Solar minima, Earth's rotation and Little Ice Ages in the past and in the future: the North Atlantic/European case. *Global and Planetary Change* 72: 282-293

Nash, T. and 55 others 2009: Obliquity-paced Pliocene West Antarctic ice sheet oscillations. *Nature* 458: 322-328

Nicholls, R. J. and Casenave, A. 2010: Sea level rise and its impact on coastal zones. *Science* 328: 1517-1520

Nick, F. M., Viell, A., Howat, I. M. and Joughin, I. 2009: Large scale changes in Greenland outlet glacier dynamics triggered at the terminus. *Nature Geoscience* 2: 110-114

O'Donnell, R., Lewis, N., McIntyre, S. and Condon, J. 2011: Improved methods for PCA-based reconstructions:case study using the Steig et al. (2009) Antarctic temperatiure reconstruction. *Journal of Climate* doi:10.1175/2010JCL13656.1

Oke, T. R. 1973: City size and the urban heat island. *Atmospheric Environment* 7: 769-779

Playford, P. 2010: The climate debate. www.pnronline.com.au/article,php/105/944

Remer, L. A., Kleidman, R. G., Levy, R. C., Kaufman, Y. J., Tanre, D., Mattoo, S., Martins, J. V., Ichoku, C., Koren, I., Yu, H. B. and Holben, B. N. 2008: Global aerosol climatology from the MODIS satellite sensors. *Journal of Geophysical Research* 113 (D14), D14SO7

Robinson, A. B., Robinson, N. E. and Soon, W. 2007: Environmental effects of increased atmospheric carbon dioxide. *Journal of American Physicians and Surgeons* 12: 79-90.

Singer, S. F. 2011: Lack of consistency between modelled and observed temperature trends. *Energy and Environment* 22: 375-406

Soares, P. C. 2010: Warming power of CO_2 and H_2O: Correlations with temperature changes. *International Journal of Geosciences* 1: 102-112

Soon, W. 2009: Solar-Arctic mediated climate variation on multidecadal to centennial timescales: Emprircal evidence, mechanistic explanation, and testable consequences. *Physical Geography* 30: 144-148

Spencer, R. W. and Braswell, W. D. 2011: On the misdiagnosis of climate feedbacks from variations in the Earth's radiant energy balance. *Remote Sensing* 3: 1603-1613

Tans, P. 2009: An accounting of the observed increase in oceanic and atmospheric CO_2 and an outlook for the future. *Oceanography* 22: 26-35.

Trenberth, K. E. and Fasullo, J. T. 2010: Tracking Earth's energy. *Science* 328: 316-317

Tremberth, K. E., Fasullo, J. T., O'Dell, C. and Wong, T. 2010: Relationships between tropical sea surface temperature and top of atmosphere radiation. *Geophysical Research Letters* 37: L03702

Trouet, V., Esper, J., Graham, N. E., Baker, A., Scourse, J. D. and Frank, D. C. 2009: Persistent positive North Atlantic Oscillation mode dominated by the

Medieval climate anomaly. *Nature* 324: 78-80

Vermeer, M. and Rahmstorf, S. 2009: Global sea level linked to global temperature. *Proceedings of the National Academy of Sciences* 106: doi: 10.1073/pnas.0907765106,21527-21532

Vinther, B. M., Buchardt, S. L., Clausen, H. B., Dahl-Jensen, D., Johnsen, S. J., Fisher, D. A., Korener, R. M., Raynaud, D., Lipenkov, V., Andersen, K. K., Blunier, T., Rasmussen, S. O., Steffensen, J. P. and Svensson, A. M. 2009: Holocene thinning of the Greenland ice sheet. *Nature* 461: 385-388

Watson, P. J. 2011: Is there evidence yet of acceleration in mean sea level rise around mainland Australia? *Journal of Coastal Research* 27: 368-377

Webb, A. P. and Kench, P. S. 2010: The dynamic response of reef islands to sea level rise: Evidence from multi-decadal analysis of island change in the Central Pacific. *Global and Planetary Change* 72: 234-246

Winger, K., Feichter, J., Kalinowski, M. B., Sartotious, H. and Schlosser, C. 2005: A new compilation of the atmospheric [85]krypton inventories from 1945 to 2000 and its evaluation in a global transport model. *Journal of Environmental Radioactivity* 80: 183-215

www.21stcenturyscience.tech.com

www.argo.ucsd.edu

www.bishop-hill.net

www.bom.gov.au/ntc/IDO60101?IDO60101.200809.pdf

http://calderup.wordpress.com/2011/05/17/accelerator-results-on-cloud-nucleation-2/

http://cdiac.ornl.gov

http://chiefio.wordpress.com/2009/10/27/ghcn-up-north-blame-canada-comrade/

www.climateaudit.com

www.climateconversation.wordshine.co.nz

www.climatedepot.com

www.climaterealists.com

www.climatescience.org.nz

www.CO2science.org

www.cru.uea.ac.uk

www.conscious.com.au

www.davidarchibald.info/papers

www.droyspencer.com

http://en.rian.ru/papers/20091216/157260660.html

www.ersl.noaa.gov

www.fixtheclimate.com

www.galileomovement.com.au

www.geocraft.com

www.globalresearch.ca

www.heartland.org

http://icecap.us/images/iploads/CO2vsTMacRae.pdf

www.ipcc.ch

www.joannenova.com.au

www.myweb.www.edu/dbunny/research/global/glopubs.htm

www.nipccreport.org

www.noaa.gov/om/vcoop/standard.htm

www.noconsensus.org (plus Donna Laframboise's book *The delinquent teenager,* an exposure of the depths of the IPCC corruption).

www.nodc.noaa.gov

www.nofrakkingconsensus.com

www.ozclimatesense.com

www.rhysicsworld.com

www.pielkeclimatesci.com

www.quadrant.org.au

www.rossmckitrick.weebly.com

www.scienceandpublicpolicy.org

www.sciencespeak.com

REFERENCES

www.slayingthedragon.com

www.surfacestations.org

http://tidesandcurrents.noaa.gov/sltrends/sltrends.shtml

www.tynbdall.ac.uk

www.warwickhughes.com

www.wattsupwiththat.com

www.weatheraction.com

www.weatherbell.com

www.wired.com

Figures and table:

Fig. 1 www.joannenova.com.au;

Fig. 2 Hallam *et al.* (1983);

Fig. 3 Manuel *et al* (2005);

Fig. 4 http://cdiac.ornl.gov/trends/atm_methane.html;

Fig. 5 Schneider *et al.* (2006), Beck (2007) and www.co2science.org;

Figs 6-11 (incl.) http://icecap.us, http://www.niwa.co.nz and http://www.ap-pinsys.com; Table 1 from http://www.CO2science.org;

Fig. 12 National Academy of Sciences Committee on Climate Change (Younger Dryas Abrupt Climate Changes);

Fig. 13 Petit *et al.* (1999);

Fig. 14 Alley *et al.* (2004) and NOAA GISP2 ice core data;

Fig. 15 www.joannenova.com.au;

Figs. 16 and 17 IPCC 1990, Mann *et al.* 1999 and IPCC 2001;

Fig. 18 HadCRUT and UAH MSU;

Fig. 19 Raymo *et al.* (1997);

Fig. 20 www.scotese.com and Berner (2001 and 2006);

Fig. 21 Fig. 21 Willis Eschenbach (2006), Archibald (2008),

Fig. 22 NIPCC;

Fig. 23 Jawowski *et al.* (1992) and www.joannenova.com.au;

Fig. 24 see Fig. 21;

Fig. 25 Karl *et al.* (2006);

Fig. 26 www.data.giss.nasa.gov, www.warwickhughes.com and Loehle (2009);

Fig. 27 see Fig. 13;

Fig. 30 Monnon *et al.* (2004) and Jouzel *et al.* (2004);

Fig. 31 Eustatic curves;

Fig. 32 Berner (2001), www.co2science.org and Copenhagen Diagnosis (2009).

INDEX

ABOUT THE AUTHOR

PROFESSOR IAN PLIMER (The University of Adelaide) is Australia's best-known geologist. He is also Emeritus Professor of Earth Sciences at The University of Melbourne where he was Professor and Head of Earth Sciences (1991–2005) after serving at The University of Newcastle (1985–1991) as Professor and Head of Geology. He was on the staff of the University of New England, The University of New South Wales and Macquarie University. He has published more than 120 scientific papers on geology. This is his eighth book written for the general public, the best-known of which are *Telling Lies for God* (Random House), *Milos-Geologic History* (Koan), *A Short History of Planet Earth* (ABC Books) and *Heaven and Earth* (Connor Court).

He has won the Leopold von Buch Plakette (German Geological Society), Clarke Medal (Royal Society of NSW), Sir Willis Connolly Medal (Australasian Institute of Mining and Metallurgy), was elected Fellow of the Australian Academy of Technological Sciences and Engineering and was elected Honorary Fellow of the Geological Society of London. In 1995, he was Australian Humanist of the Year and later was awarded the Centenary Medal. He was Managing Editor of Mineralium Deposita, president of the SGA, president of IAGOD, president of the Australian Geoscience Council and sat on the Earth Sciences Committee of the Australian Research Council for many years. He has won the Eureka Prize for the promotion of science, the Eureka Prize for *A Short History of Planet Earth* and the Michael Daley Prize (now a Eureka Prize) for science broadcasting. He is an adviser adviser to governments and corporations and a regular broadcaster.

Professor Plimer spent much of his life in the rough and tumble of the zinc-lead-silver mining town of Broken Hill, where an integrated interdisciplinary scientific knowledge intertwined with a healthy dose of scepticism and pragmatism are necessary. His time in the outback has introduced him to those who can immediately see the weaknesses of an argument. He is Patron of Lifeline Broken Hill and the Broken Hill Geocentre. He worked for North Broken Hill Ltd, was a director of CBH Resources Ltd, and is a director of Silver City Minerals Ltd and a number of other listed companies. A new Broken Hill mineral, plimerite $ZnFe_4(PO_4)_3(OH)_5$ orthorhombic, was named in his honour in recognition of his contribution to Broken Hill geology.